항공기 정비 영어

TECHNICAL ENGLISH FOR AIRCRAFT MAINTENANCE

권오건 · 김민정 · 김형래 · 방시원 · 석흥식 · 한영동

한국항공우주기술협회

(사)한국항공우주기술협회

1992년에 창립하여 항공기술인의 친목 및 권익 보호와 교육, 출판을 통한 항공우주기술의 전파, 나아가 항공기술업무의 표준화, 기술자격관리의 선진화 및 체계화 등 국가항공기술 관련 정책 연구를 하는 국토교통부 승인 비영리단체이다.

집필위원

권오건
한국항공우주기술협회 교육위원장
아세아항공직업전문학교 학장

김민정
극동대학교 항공정비학과 교수
전) ㈜대한항공 정비본부

김형래
한국폴리텍대학 항공MRO과 교수
전) 한국항공우주산업(주)

방시원
아세아항공직업전문학교 교양교수
전) YBM 토익 강사

석흥식
㈜샤프테크닉스케이 수석부장
전) ㈜대한항공 정비본부

한영동
한서대학교 항공기술교육원 교수
전) 아시아나항공(주) 정비본부

항공기 정비 영어
TECHNICAL ENGLISH FOR AIRCRAFT MAINTENANCE

발행일 2024년 11월 15일
저자 (사)한국항공우주기술협회 · 권오건 · 김민정 · 김형래 · 방시원 · 석흥식 · 한영동
펴낸이 이정수
책임 편집 최민서 · 신지항
펴낸곳 연경문화사
등록 1-995호
주소 서울시 강서구 양천로 551-24 한화비즈메트로 2차 807호
대표전화 02-332-3923
팩시밀리 02-332-3928
이메일 ykmedia@naver.com
값 28,000원
ISBN 978-89-8298-219-4 (93550)

항공기 정비 영어

TECHNICAL ENGLISH FOR AIRCRAFT MAINTENANCE

권오건 · 김민정 · 김형래 · 방시원 · 석흥식 · 한영동

한국항공우주기술협회

발간사

공자께서 논어 계씨편 제 13장에 이런 말씀을 하였다고 합니다. "不學詩 無以言(불학시 무이언), 시를 배우지 않으면 말을 할 수 없다." 물론 시를 배우지 않아도 말을 할 수 있지만 여기서는 말다운 말, 상대방을 설득할 수 있는 말, 의사가 확실하게 전달될 수 있는 말을 의미한다고 할 수 있습니다. 이 말을 이렇게 바꾸어 보면 어떨까요? "不學STE 無以航整, 항공기술영어를 배우지 않으면 항공 정비를 할 수 없다."

항공 분야에서는 우리 모두가 잘 아는 바와 같이 모든 관련 매뉴얼과 기타 문서들은 모두 영어로 되어 있습니다. 물론 일부 문서들은 한국어로 번역이 되어 있으나 모든 매뉴얼과 관련 문서들을 적시에 모두 번역하여 제공하는 것은 불가능합니다.

우리는 어릴 때부터 영어를 배워 왔으나 항공기 분야에서 사용하는 단어는 일반적으로 사용하는 단어와 같더라도 의미가 다를 수 있고 같은 단어라 할지라도 여러 의미가 있고 불명확한 의미가 있을 수 있습니다. 단어의 의미를 정확하게 표현하지 못하면 전달하고자 하는 의미가 전달되지 못하게 되고 정비 작업에 심각한 오류가 발생할 수 있습니다. 이는 항공기 안전에도 심각한 저해 요인이 발생할 수 있음을 의미합니다. 그만큼 언어 소통은 매우 중요하다고 할 수 있습니다.

한국항공기술협회에서는 2022년에 유럽항공우주산업협회(ASD)가 발간한 ASD-STE100을 보완하여 편찬한 '항공기술영어(Aviation Technical English)'를 발간하였습니다. 이 교재는 항공기술영어를 강의하시는 분들을 위한 교재였으며 여러 차례 항공사와 학교 관계자들과 세미나를 가져본 결과 모든 참석자들이 본 교재의 필요성에 대해 매우 공감하며 크게 환영하는 분위기임을 확인하였습니다.

이 교재는 항공기 정비사를 지망하는 분들과 현업에 종사하지만 항공기술영어에 자신감이 부족하다고 느꼈던 분들에게 짧은 기간 안에 자신감을 회복할 수 있도록 도움을 줄 수 있다고 확신합니다. 이 교재를 출간하기 위하여 불철주야 각고의 노력을 아끼지 않으신 권오건 위원장님 그리고 김민정, 김형래, 방시원, 석흥식, 한영동 위원님들의 헌신을 치하하며 깊이 감사를 드립니다. 이 교재가 한국의 항공 안전의 향상과 발전에 크게 기여하게 되리라 믿습니다.

2024. 11.

(사)한국항공우주기술협회장 김 세 한

추천사

전 세계가 항공운송산업의 발달로 급속도로 이웃과 같이 좁혀지고 있다.

이러한 환경의 변화로 영어의 쓰임새가 더욱 중요하게 되면서 만국의 공용어로 위상을 한껏 드높이고 있기도 하다.

그래서인지 우리들 모두가 선택의 여지가 없이 영어가 필수적인 언어로 자리를 잡아 어렸을 때부터 익히도록 강요되고 있다고 해도 과언이 아니다.

항공산업이 미국을 중심으로 성장하였기 때문에 당연히 영어가 항공기술의 전달수단으로 절대적인 역할을 하지만 전문가로 성장하기 위해서는 고유 전문용어에 대한 지식을 기본적으로 갖추어야 한다.

이러한 현실에서 국제적으로 규정한 지침에 따라 항공기술에 대한 전문영어도서가 출판된 것은 항공산업분야의 안전한 발전을 위하여 시의적절하다고 생각한다.

항공관련기술 양성학교뿐만 아니라 항공산업계에서도 일반영어가 아니라 기술영어에 대한 실용적인 학습을 통하여 궁극적으로 국제적인 항공산업안전의 토양을 구축할 것을 강력히 추천하는 바이다.

2024. 11.

전) (사)한국항공우주기술협회장 이 상 희

Table of Contents

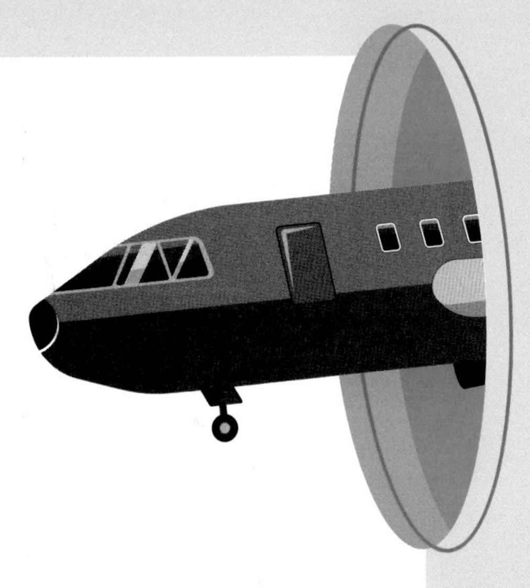

PART 1
Elements of English Sentences

영어 문장의 구성 요소

Part 1에서는 영어 입문자들이 항공기 기술 도서의 독해 능력을 높이는 훈련을 제공한다. 이 과정은 영어 문장의 성분과 요소의 위치를 이해하는 데 중점을 둔다. 실제 항공기 매뉴얼에서 발췌한 예문을 통해 학습자들이 문장 스타일에 적응하고, 실무에 바로 적용할 수 있는 독해 능력을 키울 수 있도록 도와준다.

PART 1.1 문장의 형식

영어 문장은 다섯 개의 형식으로 구성된다. 각 형식은 문장 성분과 그 성분을 채우는 구성 요소로 이루어진다. 문장 성분에는 주어, 동사, 목적어, 보이가 있으며, 이를 채우는 구성 요소로는 8품사 중 명사, 대명사, 동사, 형용사, 부사가 있다.

1) 문장의 성분 및 분석 기호

형식	문장 성분							
1형식	S	주어	1V	완전자동사				
2형식	S	주어	2V	불완전자동사	SC	주격보어		
3형식	S	주어	3V	완전타동사	O	목적어		
4형식	S	주어	4V	수여동사	IO	간접목적어	DO	직접목적어
5형식	S	주어	5V	불완전타동사	O	목적어	OC	목적격보어

2) 기호의 철자와 뜻

기호	용어 및 뜻	기호	용어 및 뜻
S	**S**ubject 주어	IO	**I**ndirect **O**bject 간접목적어
V	**V**erb 동사	DO	**D**irect **O**bject 직접목적어
SC	**S**ubject **C**omplement 주격보어	OC	**O**bject **C**omplement 목적격보어
O	**O**bject 목적어		

1형식(완전자동사)

1.2.1 개념정리

1) 1형식 문장의 구조와 이해

1형식 문장의 필수 성분은 주어와 동사이다. 주어(S)는 명사, 명사구, 명사절로, 동사는 완전 자동사(1V)로 구성되어 문장을 완성한다.

2) 1형식 문장의 예문

1형식 문장은 필수 성분과 그 나머지 부분을 구분할 수 있어야 하며, 나머지 부분은 수식어구 역할을 하는 부사, 전치사구, 부정사구 등이 있다.

❶ Air (S) goes (1V) to the air motor. — 공기는 공기 모터로 들어간다.

❷ The oil (S) returns (1V) to the tank. — 오일은 탱크로 되돌아간다.

❸ One specific signal (S) goes (1V) to the airplane. — 하나의 특정 신호가 항공기로 전송된다.

❹ This (S) can occur (1V) under different conditions. — 이는 다양한 조건에서 발생할 수 있다.

❺ The pressure release valve (S) opens (1V) at 15 psig. — 압력 방출 밸브는 15 psig에서 열린다.

❻ A dirty air filter (S) can result (1V) in slow operation of the start valve. — 공기 필터가 더러우면 스타트 밸브의 작동이 느려질 수 있다.

1.2.2 문장분석

완전자동사는 동사만 사용하는 단일형과 동사에 전치사를 수반하여 그 뒤에 전치사구로 이어지는 전치사 수반형의 두 가지 형태가 있다.

1) S + 1V(완전자동사)

주어와 완전자동사는 1형식 필수 문장 성분이다. 이들만으로도 1형식 문장이 완성될 수 있으며, 수식어구를 추가하면 보다 풍부한 의미를 지닌 1형식 문장이 될 수 있다.

❶ The oxygen mask (S) drops (1V) automatically.	산소마스크는 자동으로 떨어진다.
❷ The autopilot (S) engages (1V) during long flight.	자동 조종 장치는 장거리 비행 중에 작동된다.
❸ Hydraulic fluid (S) circulates (1V) through the system.	유압액(작동유)은 시스템을 따라 순환한다.
❹ The turbine (S) spins (1V) at high speed in operation.	터빈은 작동 중에 고속으로 회전한다.
❺ The opposite (S) occurs (1V) during power reductions.	출력 감소 시에는 반대 현상이 발생한다.
❻ The pump (S) did not operate (1V) during maintenance.	유지보수 중에 펌프가 작동하지 않았다.
❼ The hydraulic system (S) functions (1V) under pressure.	유압 시스템은 압력하에서 작동한다.
❽ Engine vibration (S) shows (1V) on a vertical analog scale.	엔진 진동은 수직 아날로그 스케일로 표시된다.
❾ The interlock actuator motor (S) retracts (1V) under pressure.	상호 잠금 작동기 모터는 압력이 가해지면 접혀진다.

2) S + 1V(완전자동사) + 전치사구

몇몇 완전자동사는 전치사를 수반하여 전치사구를 이끈다. 전치사구가 없으면 의미가 완전하지 않으므로 전치사를 포함시켜야 한다. 이는 전치사가 필요 없는 완전자동사와 다르다.

❶ The radar scans for obstacles. (S: The radar / 1V: scans)
레이더는 장애물을 탐지한다.

❷ The engine starts with a loud noise. (S: The engine / 1V: starts)
엔진은 큰 소음과 함께 시작된다.

❸ This results in a strong, stiff structure. (S: This / 1V: results)
이는 강하고 단단한 구조를 만들어 낸다.

❹ The warning system alerts to any issues. (S: The warning system / 1V: alerts)
경고장치는 모든 이슈에 대해 경고한다.

❺ The door will not seat in the door frame. (S: The door / 1V: will not seat)
문이 문틀에 맞지 않을 것이다.

❻ The landing gear extends from the fuselage. (S: The landing gear / 1V: extends)
착륙 장치는 동체로부터 펼쳐진다.

❼ The aircraft complies with safety regulations. (S: The aircraft / 1V: complies)
항공기는 안전 규정을 준수한다.

❽ The stator bolts to the gearbox over the rotor. (S: The stator / 1V: bolts)
스테이터는 로터 위에 있는 기어박스에 볼트로 고정된다.

❾ The fan speed sensor mounts to the fan frame. (S: The fan speed sensor / 1V: mounts)
팬 속도 센서는 팬 프레임에 장착된다.

❿ The blades range from 4.75 inches to 6 inches. (S: The blades / 1V: range)
깃(브레이드)의 길이는 4.75 인치에서 6인치 범위에 있다.

PART 1.2 연습문제

Answer Keys p. 41

1 문장분석

아래 문장에서 주어와 1형식 완전자동사에 밑줄을 긋고, 주어진 기호에 따라 표시하시오.

기호 ›› S(주어), 1V(완전자동사)

01 A tube for PS3 goes to the ECU.

02 The seal must adhere correctly.

03 You can progress with the test.

04 The relief valve opens at 15 psig.

05 This occurs during a manual start.

06 A warning flag appears in the window.

07 Each antenna consists of three parts.

2 문장완성

아래 한글 문장의 의미에 맞게 제시된 표현들을 적절히 조합하여 올바른 문장을 완성하시오.

01 착륙 장치는 항공기에 볼트로 고정된다.
to the aircraft / bolts / the landing gear

____

02 당신은 이 지침들을 준수해야 한다.
these instructions / comply with / you / must

____

03 귀환 연료는 바이패스 리턴으로 간다.
goes to / the return fuel / the bypass return

____

04 이것은 다양한 조건에서 발생할 수 있다.
under different conditions / occur / this / can

____

05 그것은 클램프로 냉각 매니폴드에 장착된다.
mounts / to the cooling manifold / by a clamp / it

____

06 올바른 접착은 표면의 평탄도에 달려 있다.
on the surface flatness / a correct bond / depends

____

07 엔진 오일 시스템 지시는 EICAS에 표시된다.
on EICAS / show / the engine oil system indications

____

PART 1.3 2형식(불완전자동사)

1.3.1 개념정리

1) 2형식 문장의 구조와 이해

S	주어 명사·구·절	+	2V	불완전 자동사	+	SC	주격보어 형용사, 과거분사, 전치사구 명사·구·절, to부정사 명사적용법

2형식 문장의 필수 성분은 주어, 동사, 보어이다. 주어(S)는 명사, 명사구, 명사절로, 동사는 불완전자동사(2V)로, 주격보어(SC)는 주어를 꾸며주는 형용사, 과거분사, 전치사구 및 명사, 명사구, 명사절 그리고 to부정사 명사적용법으로 구성되어 문장을 완성한다.

2) 2형식 문장의 예문

2형식 문장에서 주격보어는 주어의 상태나 정체를 나타낸다. 문장은 필수 성분과 그 나머지 부분을 구분할 수 있어야 하며, 나머지 부분은 수식어구 역할을 하는 부사, 전치사구, 분사구 등이 있다.

❶ The unlocked indication(S) is(2V) clearly visible(SC).
잠금 해제 표시가 명확하게 보인다.

❷ The engine noise(S) becomes(2V) louder(SC) at full throttle.
엔진 소음이 전속력에서 더 커진다.

❸ Solvent(S) is(2V) highly inflammable(SC) in a confined space.
솔벤트는 밀폐 공간에서 높은 인화성이 있다.

❹ The cabin pressure(S) stays(2V) stable(SC) at cruising altitude.
객실 압력은 순항 고도에서 안정적으로 유지된다.

❺ Directional control valve(S) is(2V) opened(SC) by manual override.
방향제어밸브는 수동적인 작동에 의해 열린다.

⑥ The turbine blades are visible through the inspection panel. 터빈 블레이드가 점검 패널을 통해 보인다.
S / 2V / SC

⑦ The engine remains operational even during heavy turbulence. 엔진은 심한 난기류 동안에도 작동을 유지한다.
S / 2V / SC

⑧ These two cautions must be observed at all times during operation. 이 두 가지 주의 사항은 작동 중 항상 준수되어야 한다.
S / 2V / SC

1.3.2 문장분석

주격보어는 주어를 정의하거나 상태를 나타내는 역할을 하며, 주로 형용사, 명사, 명사구, 과거분사, 전치사구, to부정사의 명사적 용법의 형태를 취한다.

1) S + 2V + SC(형용사)

주격보어로 형용사가 사용될 때, 이는 주어의 상태나 본질적인 성질을 설명한다.

① A value of 2 mm is acceptable. 2 mm 값이 허용된다.
S / 2V / SC(상태)

② This procedure is no longer applicable. 이 절차는 더 이상 적용되지 않는다.
S / 2V / SC(상태)

③ Closure motion should be smooth. 닫힘 동작은 부드러워야 한다.
S / 2V / SC(성질)

④ The actuator response is immediate. 액추에이터의 반응이 즉각적이다.
S / 2V / SC(성질)

2) S + 2V + SC(명사·구)

주격보어에 명사나 명사구가 올 때, 이는 주어를 정의하거나 정체를 나타낸다.

❶ Engine oil is the hydraulic medium.
S 2V SC(정의)

엔진 오일은 유압 매체이다.

❷ The center drive unit is an air motor.
S 2V SC(정의)

중앙 구동 장치는 공기 모터이다.

❸ Wash primer is a dangerous product.
S 2V SC(정체)

워시 프라이머는 위험한 제품이다.

❹ The landing gear is a critical component.
S 2V SC(정체)

착륙 장치는 중요한 구성품이다.

3) S + 2V + SC(과거분사)

주격보어에 과거분사가 올 때, 이는 주어의 상태를 설명한다. 과거분사는 동사에서 파생되었지만 품사가 형용사이므로 "~된"으로 해석되며, 주로 be동사와 함께 수동태 구조로 사용된다.

❶ The hydraulic fluid is filtered.
S 2V SC

작동유는 필터링 된다.

❷ This fan air flow is unregulated.
S 2V SC

이 팬 공기 흐름은 조절되지 않는다.

❸ Only one coil is energized at a time.
S 2V SC

한 번에 하나의 코일에만 전원이 공급된다.

❹ Each drain valve is labeled for identification.
S 2V SC

각 배출 밸브는 식별을 위해 라벨이 부착되어 있다.

4) S + 2V + SC(전치사구)

주격보어로 전치사구가 사용될 때, 이는 주로 주어의 위치, 방향, 상태 또는 조건을 설명한다.

❶ One switch module is on each CDU.
S / 2V / SC(위치)
각 CDU에는 하나의 스위치 모듈이 있다.

❷ Access is through the right fan cowl.
S / 2V / SC(방향)
접근은 오른쪽 팬 카울을 통해 이루어진다.

❸ The circuit breaker is in need of replacement.
S / 2V / SC(상태)
서킷 브레이커는 교체가 필요하다.

❹ This procedure is for ground maintenance use only.
S / 2V / SC(조건)
이 절차는 지상 정비 용도로만 사용된다.

5) S + 2V + SC(to부정사의 명사적 용법)

주격보어에 to부정사의 명사적 용법이 사용될 때, 이는 주로 주어의 계획, 본질, 절차, 목표, 역할, 목적을 설명한다.

❶ The plan is to upgrade the avionics.
S / 2V / SC(계획)
계획은 항공 전자 장비를 업그레이드하는 것이다.

❷ The turbine's function is to generate thrust.
S / 2V / SC(본질)
터빈의 본질적 기능은 추력을 생성하는 것이다.

❸ The procedure is to check electrical connections.
S / 2V / SC(절차)
절차는 전기 연결을 점검하는 것이다.

❹ The goal is to improve flight control responsiveness.
S / 2V / SC(목표)
목표는 비행 제어의 반응성을 향상시키는 것이다.

❺ The main role of the fuselage is to support the aircraft.
S / 2V / SC(역할)
동체의 주요 역할은 항공기를 지지하는 것이다.

❻ The purpose of the fuel pump is to supply fuel to the engine.
S / 2V / SC(목적)
연료 펌프의 목적은 엔진에 연료를 공급하는 것이다.

PART 1.3 연습문제

Answer Keys p. 42

1 문장분석

아래 문장에서 주어, 2형식 불완전자동사, 그리고 주격보어에 밑줄을 긋고, 주어진 기호에 따라 표시하시오.

기호 ›› S(주어), 2V(불완전자동사), SC(주격보어)

❶ The fuel line is under the wing near the fuselage.

❷ The brake fluid appears clear after the inspection.

❸ The engine remains efficient even with extensive use.

❹ The emergency lights are crucial during power outages.

❺ The landing gear remains stable even in rough weather.

❻ The oxygen tank is in the rear compartment of the cabin.

❼ The hydraulic fluid looks clean after the filter replacement.

2 문장완성

아래 한글 문장의 의미에 맞게 제시된 표현들을 적절히 조합하여 올바른 문장을 완성하시오.

❶ 계획은 연료 필터를 교체하는 것이다.
to replace / is / the plan / the fuel filter

____

❷ 연료 탱크는 급유 후에 가득 차 있다.
after refueling / the fuel tank / remains / full

____

❸ 정비 기록은 제어판 위에 있다.
is / the control panel / on / the maintenance log

____

❹ 착륙 장치는 안전한 착륙을 위한 필수적인 부분이다.
for safe landings / the landing gear / is / a vital part

____

❺ 조종석 조명은 야간 운용을 위해 밝다.
the cockpit lights / are / for night operations / bright

____

❻ 항공기는 매 비행 전에 철저히 점검된다.
thoroughly / before each flight / is / inspected / the aircraft

____

❼ 비상등은 전원 고장 시 작동된다.
the emergency lights / are / during power failures / activated

____

3형식(완전타동사)

1.4.1 개념정리

1) 3형식 문장의 구조와 이해

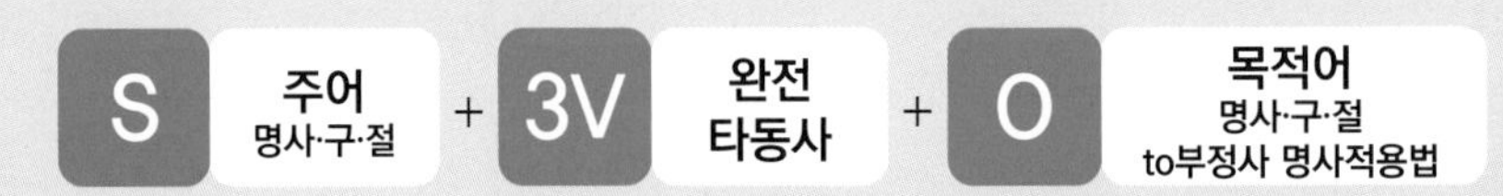

3형식 문장의 필수 성분은 주어, 동사, 목적어이다. 주어(S)는 명사, 명사구, 명사절로, 동사는 완전타동사(3V)로, 목적어(O)는 명사, 명사구, 명사절로 구성되어 문장을 완성한다.

2) 3형식 문장의 예문

3형식 문장은 주어, 동사, 목적어로 구성된다. 주어는 동작을 수행하는 사람이나 사물, 동사는 그 동작, 목적어는 그 동작의 대상을 나타낸다. 문장에서 필수 성분인 주어, 동사, 목적어와 부사, 전치사구, 부정사구, 분사구 등의 수식어구를 구분할 수 있어야 한다.

❶ This (S) completes (3V) the procedure (O) successfully.

이로써 성공적으로 절차가 완료된다.

❷ Do not apply (3V) excess adhesive (O) on the gasket.

가스켓에 과도한 접착제를 바르지 않는다.

❸ Ensure (3V) that covers are in place (O) before starting the engine.

엔진을 시작하기 전에 덮개가 제자리에 있는지 확인한다.

❹ Do not use (3V) over 100 pounds force (O) to push latch handle closed.

래치 핸들을 밀어 닫을 때 100파운드 이상의 힘을 사용하지 않는다.

❺ The batteries (S) generate (3V) 28 VDC (O) under normal operating conditions.

배터리는 정상 작동 조건에서 28볼트의 직류전압을 생성한다.

❻ Regulate the electrical current carefully to avoid any short circuits.
3V / O

어떤 단락도 피하기 위해 전류를 신중하게 조절한다.

❼ Flush the tubes with fresh water thoroughly to remove all contaminants.
3V / O

모든 오염 물질을 제거하기 위해 튜브를 깨끗한 물로 철저히 세척한다.

❽ Check that the other microswitch is operational during the pre-flight inspection.
3V / O

비행 전 점검 동안 다른 마이크로스위치가 작동하는지 확인한다.

1.4.2 문장분석

완전타동사 뒤에 오는 목적어로는 주로 명사, 명사구, to부정사의 명사적 용법, 또는 명사절이 사용되어 3형식 문장을 완성한다. 주어가 생략된 명령문에서는 'You'가 문장 앞에 있다고 간주하여 문장을 분석한다.

1) S + 3V + O(명사·구)

목적어로 명사나 명사구가 올 때, 이는 주어가 행동하는 동작의 대상을 설명한다. 해석은 주로 "~을, ~를"로 한다.

❶ This section describes the procedures in detail.
S / 3V / O

이 섹션은 절차를 자세히 설명한다.

❷ A carbon seal prevents fuel leakage during operation.
S / 3V / O

카본 씰은 작동 중 연료 누출을 방지한다.

❸ This technique gives the best results when applied correctly.
S / 3V / O

이 기술은 올바르게 적용될 때 최상의 결과를 준다.

❹ The fuel flow transmitter measures the fuel flow rate accurately.
S / 3V / O

연료 유량 송신기는 연료 유량을 정확하게 측정한다.

2) (You) + 3V + O

항공기 정비 매뉴얼에서는 작업 지시가 많기 때문에 명령문이 자주 사용된다. 이때 주어는 생략되고 'You'로 암시된다. 명령문은 보통 3형식 구조를 따르며, 동사 뒤에 목적어를 붙여 문장을 완성한다.

❶ Take(3V) a sample(O) of hydraulic fluid for analysis.
분석을 위해 작동유 샘플을 채취해야 한다.

❷ Hold(3V) the cylinder(O) firmly to prevent movement.
움직임을 방지하기 위해 실린더를 단단히 잡아야 한다.

❸ Depress and release(3V) the button(O) to ensure proper function.
제대로 작동하는지 확인하기 위해 버튼을 눌렀다가 떼야한다.

❹ Perform(3V) the leak test(O) according to the manual instructions.
매뉴얼 지침에 따라 누출 검사를 수행해야 한다.

3) (You) + 3V + that + S + V

3형식 명령문에서 목적어로 that 명사절을 사용할 수 있다. 이때, 3형식 동사는 특정 작업이나 지시를 나타내며, that 명사절은 이 지시 내용을 자세히 설명한다.

❶ Ensure(3V) that covers are in place(O) before starting the engine.
엔진을 시작하기 전에 덮개가 제자리에 있는지 확인한다.

❷ Verify(3V) that the translating cowl is fully stowed(O) after landing.
착륙 후 이동식 카울이 완전히 닫혔는지 확인한다.

❸ Secure(3V) that the pump is inactive(O) before performing maintenance.
정비를 수행하기 전에 펌프가 비활성 상태인지 확인한다.

❹ Check(3V) that the other microswitch is operational(O) during pre-flight inspection.
비행 전 점검 동안 다른 마이크로스위치가 작동하는지 점검한다.

4) (You) + do not + 3V + O(명사·구)

부정 명령문에서는 'do not' 뒤에 3형식 동사와 목적어를 사용하여 특정 행동을 금지한다. 이 구조는 명령문에서 금지된 행동을 명확히 전달한다.

❶ Do not put (3V) excess weight (O) on the trolley.

트롤리에 초과 무게를 싣지 않는다.

❷ Do not enter (3V) the engine test area (O) without approval.

승인 없이 엔진 시험 구역에 들어가지 않는다.

❸ Do not check (3V) the oil (O) for at least five minutes after shutdown.

엔진 정지 후 최소 5분 동안은 오일을 점검하지 않는다.

❹ Do not apply (3V) extreme loads (O) to the structure to avoid damage.

손상을 피하기 위해 구조물에 과도한 하중을 가하지 않는다.

PART 1.4 연습문제

Answer Keys p. 43

1 문장분석

아래 문장에서 주어, 3형식 완전타동사와 목적어에 밑줄을 긋고, 주어진 기호에 따라 표시하시오.

기호 ›› S(주어), 3V(완전자동사), O(목적어)

❶ Monitor the fuel efficiency regularly.

❷ Adjust the flight control surfaces precisely.

❸ The turbine delivers constant rotational energy.

❹ Do not operate the equipment beyond its capacity.

❺ Stabilize the hydraulic pressure within the acceptable range.

❻ The generator produces 400 Hz AC power for onboard systems.

❼ Check that all connections are secure before starting the engine.

2 문장완성

아래 한글 문장의 의미에 맞게 제시된 표현들을 적절히 조합하여 올바른 문장을 완성하시오.

❶ 주요 수리 후 배선이 온전한지 확인한다.
verify / is intact / after every major repair / that the wiring

____

❷ 누설이나 손상이 있는지 연료 라인을 철저히 점검한다.
for any leaks or damage / thoroughly / the fuel line / inspect

____

❸ 지시기는 디스플레이 패널에 현재 상태를 보여준다.
on the display panel / the indicator / the current status / shows

____

❹ 체크리스트는 매 비행 전에 필요한 점검을 명시한다.
before every flight / the checklist / the required checks / specifies

____

❺ 엔진을 끈 직후에 뜨거운 엔진 부품을 만지지 않는다.
after shutdown / do not touch / the hot engine parts / immediately

____

❻ 추가 문제를 방지하기 위해 손상된 부품을 즉시 교체한다.
the damaged component / to avoid further issues / immediately / replace

____

❼ 절대적으로 필요하지 않은 한, 작동 중에 패널을 제거하지 않는다.
unless absolutely necessary / do not / the panel / during operation / remove

____

PART 1.5 4형식(수여동사)

1.5.1 개념정리

1) 4형식 문장의 구조와 이해

4형식 문장의 필수 성분은 주어, 동사, 간접목적어, 직접목적어이다. 주어(S)는 명사, 명사구, 명사절로, 동사는 수여동사(4V)로, 간접목적어(IO)는 사람 또는 사물로, 직접목적어(DO)는 사물 또는 정보로 구성되어 문장을 완성한다.

2) 4형식 문장의 예문

4형식 문장에서는 간접 목적어와 직접 목적어를 구분할 필요가 있다. 어순은 반드시 "~에게 ~을 수여한다" 식으로 파악해야 한다.

❶ The left actuators (S) send (4V) channel A (IO) feedback signals (DO).
왼쪽 액추에이터는 채널 A에 피드백 신호를 보낸다.

❷ The sensor (S) sends (4V) the display (IO) the temperature reading (DO).
센서는 디스플레이에 온도 측정값을 보낸다.

❸ The indicator (S) shows (4V) the mechanic (IO) the current temperature (DO).
지시기는 정비사에게 현재 온도를 보여준다.

❹ The control system (S) gives (4V) the valve (IO) the correct electrical signal (DO).
제어 시스템은 밸브에 올바른 전기 신호를 준다.

❺ The control unit (S) tells (4V) the operator (IO) that the pressure is too high (DO).
제어 장치는 운영자에게 압력이 너무 높다고 말한다.

❻ The EEC (S) informs (4V) EICAS (IO) that it is operating in the reversionary mode (DO).
EEC는 EICAS에게 리버전 모드로 작동 중이라고 알린다.

❼ The alert system notifies the pilot that there is a malfunction in the engine.
S 4V IO DO
경고 시스템은 파일럿에게 엔진에 고장이 있다고 알린다.

❽ The TMC disconnect signal informs the ECU that the autothrottle is disengaged.
S 4V IO DO
TMC 분리 신호는 ECU에게 오토스로틀이 해제되었다고 알린다.

3) 4형식 → 3형식 문장전환의 규칙

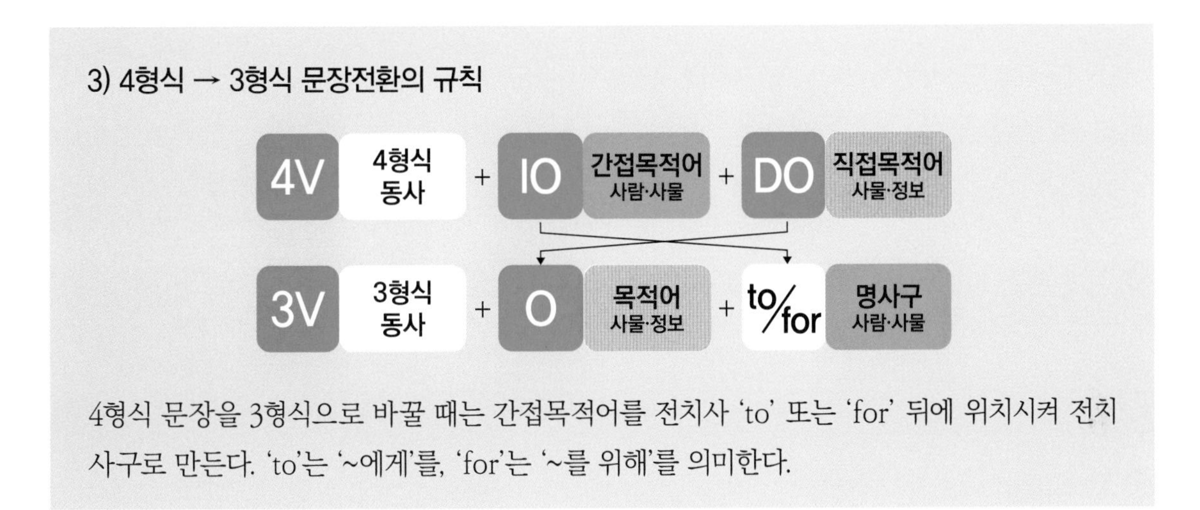

4형식 문장을 3형식으로 바꿀 때는 간접목적어를 전치사 'to' 또는 'for' 뒤에 위치시켜 전치사구로 만든다. 'to'는 '~에게'를, 'for'는 '~를 위해'를 의미한다.

❶ The controller sends the engine the start signal.
S 4V IO DO
The controller sends the start signal to the engine.
S 3V O
제어 장치는 시동 신호를 엔진에게 보낸다.

❷ The pump gives the system the required pressure.
S 4V IO DO
The pump gives the required pressure to the system.
S 3V O
펌프는 필요한 압력을 시스템에게 준다.

❸ The onboard computer sends the technician a maintenance reminder.
S 4V IO DO
The onboard computer sends a maintenance reminder to the technician.
S 3V O
탑재된 컴퓨터는 기술자에게 정비 알림을 보낸다.

❹ This procedure gives the technician instructions on how to attach the pylon.
S 4V IO DO
This procedure gives instructions to the technician on how to attach the pylon.
S 3V O
이 절차는 기술자에게 파일런을 부착하는 방법에 대한 지침을 제공한다.

1.5.2 문장분석

4형식 문장의 간접목적어에는 사람이나 사물의 명사구가 사용되고, 직접목적어에는 사물의 명사구 또는 "that" 명사절을 이용한 정보가 위치한다.

1) S + 4V + IO(사람) + DO

수여동사 뒤의 간접목적어는 주로 사람 명사로, 수여대상을 나타내며 "~에게"로 해석한다. 그 뒤에 나오는 직접목적어는 사물을 나타내며 "~을, ~를"로 해석한다.

❶ The alert / gives / the engineer / a critical update.
S / 4V / IO / DO

경고는 엔지니어에게 중요한 업데이트를 준다.

❷ The system / sends / the technician / a warning signal.
S / 4V / IO / DO

시스템은 기술자에게 경고 신호를 보낸다.

❸ The indicator / shows / the operator / the current status.
S / 4V / IO / DO

지시기는 운영자에게 현재 상태를 보여준다.

❹ The manual / offers / the technician / detailed instructions.
S / 4V / IO / DO

매뉴얼은 기술자에게 자세한 지침을 제공한다.

2) S + 4V + IO(사물) + DO

수여동사 뒤의 간접목적어가 사물일 때, 이를 "~에"로 해석하고, 직접목적어는 "~을, ~를"로 해석한다.

❶ The sensor / offers / the circuit / the data it needs.
S / 4V / IO / DO

센서는 회로에 필요한 데이터를 제공한다.

❷ The unit / shows / the device / the required settings.
S / 4V / IO / DO

유닛은 장치에 필요한 설정을 보여준다.

❸ The controller / gives / the engine / the ignition signal.
S / 4V / IO / DO

컨트롤러는 엔진에 점화 신호를 준다.

❹ The pump / sends / the system / the necessary pressure.
S / 4V / IO / DO

펌프는 시스템에 필요한 압력을 보낸다.

3) S + 4V + IO + DO(사물)

4형식 문장에서 수여동사의 뒤에 간접목적어가 사람일 때는 "~에게"로 해석하고, 간접목적어가 사물일 때는 "~에"로 해석한다. 직접목적어는 주로 사물로 사용된다.

❶ The alert system(S) sends(4V) the cockpit(IO) the warning message(DO).
경고 시스템은 조종석에 경고 메시지를 보낸다.

❷ The flight computer(S) gives(4V) the navigation system(IO) the route data(DO).
비행 컴퓨터는 항법 시스템에 경로 데이터를 준다.

❸ The diagnostic tool(S) shows(4V) the maintenance system(IO) the fault codes(DO).
진단 도구는 정비 시스템에 고장 코드를 보여준다.

❹ The training manual(S) offers(4V) the technician(IO) the troubleshooting steps(DO).
교육 매뉴얼은 기술자에게 문제 해결 단계를 제공한다.

4) S + 4V + IO(사람) + DO(정보)

몇몇 수여동사는 간접목적어로 사람명사를 주로 사용하고, 직접목적어로 정보를 전달한다. 이 경우, 직접목적어로 that 명사절을 사용하여 정보를 누구에게 전달하는지를 명확히 한다.

❶ The computer(S) tells(4V) the pilot(IO) that the route is clear(DO).
컴퓨터는 파일럿에게 경로가 명확하다고 말한다.

❷ The alert system(S) tells(4V) the crew(IO) that a door is open(DO).
경고 시스템은 승무원에게 문이 열려 있다고 알린다.

❸ The indicator(S) informs(4V) the operator(IO) that the pressure is low(DO).
지시기는 운영자에게 압력이 낮다고 알린다.

❹ The sensor(S) notifies(4V) the engineer(IO) that the temperature is rising(DO).
센서는 엔지니어에게 온도가 상승하고 있다고 알린다.

PART 1.5 연습문제

Answer Keys p. 44

1 문장분석

아래 문장에서 주어, 수여동사, 간접목적어, 직접목적어를 찾아 밑줄을 긋고 주어진 기호에 따라 표시하시오.

기호 ›› S(주어), 4V(수여동사), IO(간접목적어), DO(직접목적어)

❶ The sensor sends the system real-time data.

❷ The tool grants the motor greater precision.

❸ The module gives the apparatus stable power.

❹ The computer shows the engineer a clear diagram.

❺ The alert notifies the pilot that a main door is open.

❻ The manual gives the equipment increased reliability.

❼ The software informs the user that a new update is available.

2 문장완성

아래 한글 문장의 의미에 맞게 제시된 표현들을 적절히 조합하여 올바른 문장을 완성하시오.

❶ 경보는 운영자에게 즉각적인 경고를 준다.
gives / the operator / an immediate warning / the alert

____

❷ 시스템은 차량에 최적의 성능을 보여준다.
shows / the vehicle / the system / optimal performance

____

❸ 컨트롤러는 항공기에 향상된 안정성을 제공한다.
offers / the aircraft / enhanced stability / the controller

____

❹ 모니터는 장비에 작동 상태를 보여준다.
shows / the equipment / operational status / the monitor

____

❺ 보고서는 승무원에게 점검이 완전히 완료되었다고 알린다.
informs / is / the crew / that the check / fully complete / the report

____

❻ 제어 장치는 정비 팀에게 긴급 경보를 보낸다.
sends / the maintenance crew / an urgent alert / the control unit

____

❼ 시스템은 운영자에게 정기적인 정비가 필요하다고 말한다.
tells / the operator / that regular maintenance / needed / the system / is

____

5형식(불완전타동사)

1.6.1 개념정리

1) 5형식 문장의 구조와 이해

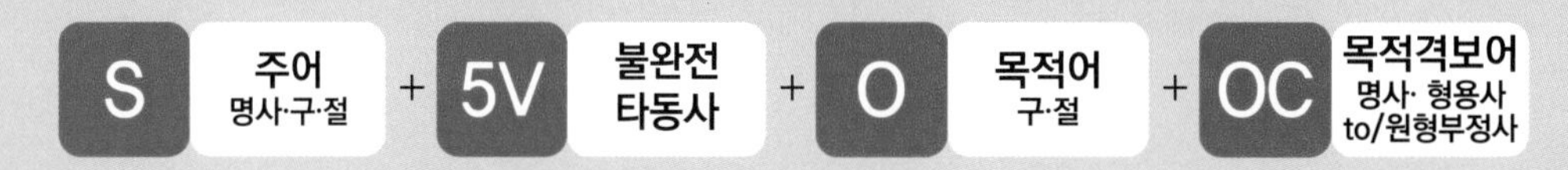

5형식 문장의 필수 성분은 주어, 동사, 목적어, 목적격보어이다. 주어(S)는 명사나 명사구로, 동사는 불완전타동사(5V)로, 목적어(O)는 명사, 명사구, 명사절로, 목적격보어(OC)는 주로 형용사, 명사나 명사구, to부정사, 원형부정사로 구성되어 문장을 완성한다.

2) 5형식 문장의 예문

5형식 문장의 목적격보어는 목적어의 상태를 설명하거나, 정체를 나타내거나, 동작을 표현하기 위해 형용사, 명사, 명사구, to부정사, 원형부정사 등을 사용하여 5형식 문장을 완성한다. 또한, 주어인 'You'를 생략하여 명령문과 부정명령문을 만들 수도 있다.

❶ Keep all personnel clear of areas.
Keep (5V) / all personnel (O) / clear (OC)
모든 인원을 해당 구역에서 떨어지게 한다.

❷ Make sure the door frame area is clear.
Make (5V) / sure (OC) / the door frame area is clear (O)
문틀 구역이 깨끗한지 확인한다.

❸ This lets the metered fuel push the shutoff valve open.
This (S) / lets (5V) / the metered fuel (O) / push the shutoff valve (OC)
이것은 측정된 연료가 차단 밸브를 밀어 열리게 한다.

❹ Warn personnel to stay away from the conveyor system.
Warn (5V) / personnel (O) / to stay away (OC)
인원에게 컨베이어 시스템에서 떨어져 있으라고 경고한다.

❺ Do not allow the cable to touch the floor.
Do not allow (5V) / the cable (O) / to touch the floor (OC)
케이블이 바닥에 닿지 않게 한다.

1.6.2 문장분석

5형식 문장의 목적격보어 형태는 대부분 형용사, 명사, to부정사구와 원형부정사구가 사용된다.

1) S + 5V + O + OC(형용사, 명사)

목적격보어에 형용사나 명사가 사용될 때, 이는 목적어의 상태나 본질적인 성질을 설명한다.

❶ The update makes the software efficient.
S 5V O OC(형)
업데이트는 소프트웨어를 효율적으로 만든다.

❷ The manual keeps the technician prepared.
S 5V O OC(형)
매뉴얼은 기술자를 준비된 상태로 유지한다.

❷ The guidelines classified the inspection a priority.
S 5V O OC(명)
지침은 그 점검을 우선순위로 분류했다.

❹ The procedure considered the action a requirement.
S 5V O OC(명)
절차는 행동을 필수사항으로 여겼다.

2) S + 5V + OC(형용사) + O(that 명사절)

5형식 문장에서 주격보어로 형용사가 먼저 오고 그 뒤에 목적어로 명사절이 오는 도치구문이 사용될 수도 있다. 이 구조는 "~을 명확히, 확실히 해두다"는 의미를 표현한다. 또한, 항공기 정비 매뉴얼에서는 'You' 주어가 암시되어 생략된 명령문이 자주 사용된다.

❶ The system makes clear that the engine is running.
S 5V OC O
시스템은 엔진이 작동 중이라는 것을 명확하게 한다.

❷ The alert makes evident that there is a malfunction.
S 5V OC O
경보는 오작동이 있다는 것을 분명히 한다.

❸ Make certain that the software is updated.
5V OC O
업데이트는 소프트웨어가 업데이트되었는지를 확실하게 하다.

❹ Makes sure that the parts are installed correctly.
5V OC O
부품이 올바르게 설치되었는지를 확실하게 하다.

3) S + 5V + O + OC(to부정사구)

목적격 보어로 to부정사구가 사용되는 이유는 5형식 동사의 의미적 성격 때문이다. 이러한 동사들은 크게 세 가지로 나눌 수 있다: 목적어에게 결과를 초래시키거나, 행동을 유발하거나, 성격이나 상태를 정의하는 경우이다.

❶ This causes the time delay to increase.
S 5V(결과초래) O OC
이것은 시간 지연을 증가시킨다.

❷ The indicator tells the operator to monitor.
S 5V(행동유발) O OC
지시기는 운영자가 모니터하도록 지시한다.

❸ The manual instructs the technician to check.
S 5V(행동유발) O OC
매뉴얼은 기술자가 점검하도록 지시한다.

❹ The technician considers the engine to be faulty.
S 5V(상태정의) O OC
기술자는 엔진이 고장 나 있다고 판단한다.

4) S + 5V(사역동사) + O + OC(원형부정사구)

'have', 'let', 'make'는 5형식 사역동사로서 목적어의 행동을 원형부정사 형태로 목적격보어에 표현한다. 예외적으로 'get'은 to 부정사를 사용한다.

❶ The supervisor had the mechanic test the avionics system.
S 5V O OC
감독관은 정비사가 항공 전자 장비 시스템을 테스트하게 했다.

❷ Two positions let the cowl be held open.
S 5V O OC
두 위치는 카울이 열린 상태로 유지되도록 한다.

❸ You can let these doors stay latched open.
S 5V O OC
이 문들이 열린 상태로 고정되도록 할 수 있다.

❹ The alert makes the operator shut down the system.
S 5V O OC
경보는 운영자가 시스템을 종료하게 한다.

5) S + 5V(사역동사) + O + OC(형용사구, 전치사구, 분사구)

사역동사 뒤의 목적격 보어 자리에 형용사구, 전치사구, 분사구와 같은 형용사 상당어구가 등장할 수 있다. 이러한 어구들은 본연의 목적격 보어처럼 목적어의 성질이나 상태를 설명한다.

❶ The diagnostic tool (S) made (5V) the powerplant (O) ready for testing (OC(형용사구)).

진단 도구는 동력장치 테스트 준비가 되게 했다.

❷ The maintenance procedure (S) had (5V) the landing gear (O) in good condition (OC(전치사구)).

정비 절차는 착륙 장치를 좋은 상태로 유지하도록 했다.

❸ The sensor (S) got (5V) the hydraulic system (O) functioning properly (OC(분사구)).

센서는 유압 시스템이 제대로 작동하게 했다.

6) (You) + 5V + O + OC(형용사 상당어구)

5형식 사역동사는 목적어의 상태에 영향을 주거나 행동을 유발시키기 위해 사용된다. 상태를 표현할 때는 형용사 상당어구가, 행동을 표현할 때는 원형부정사가 사용된다. 항공기 정비 매뉴얼의 특성상, 사역동사는 주어가 생략된 명령문으로 자주 나타난다.

❶ Let (5V) the thrust reverser (O) half open (OC(상태)).

역추진 장치를 절반만 열어 둔다.

❷ Have (5V) the flaps (O) retracted to the zero position (OC(상태)).

플랩을 0 위치로 접히게 한다.

❸ Let (5V) the fuel pump (O) remain off (OC(행동)) during maintenance.

정비 중에는 연료 펌프가 꺼진 상태로 유지되도록 한다.

❹ Have (5V) the APU (O) run (OC(행동)) for a minimum of five minutes.

APU가 최소 5분 동안 작동하게 한다.

7) (You) do not + 5V + O + OC(to부정사구, 원형부정사구)

5형식 동사가 행동을 유발시키거나 사역동사일 경우, 주어가 암시되어 생략된 명령문으로 쓰인다. 이때 목적격보어는 to부정사구나 원형부정사구로 문장이 완성된다.

❶ Do not permit the fluid to touch you.
5V / O / OC(to부정사구)

유체가 신체에 닿지 않도록 해야 한다.

❷ Do not let the battery discharge completely.
5V / O / OC(원형부정사구)

배터리가 완전히 방전되지 않도록 해야 한다.

❸ Do not permit the avionics bay to overheat.
5V / O / OC(to부정사구)

항공 전자기기실이 과열되지 않도록 허용하지 않는다.

❹ Do not allow core cowl panels to slam closed.
5V / O / OC(to부정사구)

코어 카울 패널이 쾅 닫히지 않도록 허용하지 말아야 한다.

PART 1.6 연습문제

Answer Keys p. 45

1 문장분석

아래 문장에서 주어, 5형식 불완전타동사, 목적어와 목적격보어를 찾아 밑줄을 긋고, 주어진 기호에 따라 표시하시오.

기호 ›› S(주어), 5V(불완전타동사), O(목적어), OC(목적격보어)

❶ Make sure the fuel tank is fully sealed.

❷ The procedure keeps the engine secure.

❸ Do not let the oil level drop below minimum.

❹ Do not allow the hydraulic system to overheat.

❺ The update lets the software execute the commands.

❻ The update makes sure that the software is current.

❼ The checklist asks the crew to verify all safety measures.

2 문장완성

아래 한글 문장의 의미에 맞게 제시된 표현들을 적절히 조합하여 올바른 문장을 완성하시오.

❶ 절차는 연료 라인을 깨끗하게 유지한다.
the procedure / the fuel / keeps / lines / clear

____

❷ 압력이 필요한 수준 이하로 떨어지지 않도록 한다.
below / drop / do not / the required level / the pressure / let

____

❸ 냉각 시스템이 오염되지 않도록 한다.
the cooling system / allow / contaminated / do not / to become

____

❹ 업데이트는 항법 시스템이 경로를 계산하게 한다.
the update / calculate / lets / the navigation system / the route

____

❺ 업데이트는 비상등이 작동하는지 확실하게 한다.
the update / that the emergency lights / makes / sure / functional / are

____

❻ 항공 전자기기실이 제대로 환기되고 있는지 확인한다.
make / properly ventilated/ that the avionics bay / is / sure

____

❼ 체크리스트는 정비 팀이 모든 부품을 점검하도록 요구한다.
the checklist / to inspect / the maintenance team / asks / all components

____

Answer Keys

PART 1.2 1형식(완전자동사)

1) 문장분석

❶ A tube for PS3 goes to the ECU.
S 1V
PS3 튜브는 ECU로 간다.

❷ The seal must adhere correctly.
S 1V
씰은 올바르게 밀착되어야 한다.

❸ You can progress with the test.
S 1V
테스트를 진행할 수 있다.

❹ The relief valve opens at 15 psig.
S 1V
릴리프 밸브는 15 psid에서 열린다.

❺ This occurs during a manual start.
S 1V
이것은 매뉴얼 시동 중에 발생한다.

❻ A warning flag appears in the window.
S 1V
경고 깃발이 창에 나타난다.

❼ Each antenna consists of three parts.
S 1V
각 안테나는 세 부분으로 구성되어 있다.

2) 문장완성

❶ 착륙 장치는 항공기에 볼트로 고정된다.
The landing gear bolts to the aircraft.

❷ 당신은 이 지침들을 준수해야 한다.
You must comply with these instructions.

❸ 귀환 연료는 바이패스 리턴으로 간다.
The return fuel goes to the bypass return.

❹ 이것은 다양한 조건에서 발생할 수 있다.
This can occur under different conditions.

❺ 그것은 냉각 매니폴드에 클램프로 장착된다.
It mounts to the cooling manifold by a clamp.

❻ 올바른 접착은 표면의 평탄도에 달려 있다.
A correct bond depends on the surface flatness.

❼ 엔진 오일 시스템 지시는 EICAS에 표시된다.
The engine oil system indications show on EICAS.

PART 1.3 2형식(불완전자동사)

1) 문장분석

❶ The fuel line is under the wing near the fuselage.
S / 2V / SC
연료 라인이 동체 근처 날개 아래에 있다.

❷ The brake fluid appears clear after the inspection.
S / 2V / SC
브레이크 유체가 점검 후에 깨끗하게 보인다.

❸ The engine remains efficient even with extensive use.
S / 2V / SC
엔진이 과도한 사용에도 불구하고 효율적이다.

❹ The emergency lights are crucial during power outages.
S / 2V / SC
비상등이 정전 시에 매우 중요하다.

❺ The landing gear remains stable even in rough weather.
S / 2V / SC
착륙 장치가 거친 날씨에서도 안정적이다.

❻ The oxygen tank is in the rear compartment of the cabin.
S / 2V / SC
산소 탱크가 객실 뒤쪽 구역에 있다.

❼ The hydraulic fluid looks clean after the filter replacement.
S / 2V / SC
작동유가 필터 교체 후에 깨끗하게 보인다.

2) 문장완성

❶ 계획은 연료 필터를 교체하는 것이다.
The plan is to replace the fuel filter.

❷ 연료 탱크는 주유 후에 가득 차 있다.
The fuel tank remains full after refueling.

❸ 정비 기록은 제어판 위에 있다.
The maintenance log is on the control panel.

❹ 착륙 장치는 안전한 착륙을 위한 필수적인 부분이다.
The landing gear is a vital part for safe landings.

❺ 조종석 조명은 야간 운용을 위해 밝다.
The cockpit lights are bright for night operations.

❻ 항공기는 매 비행 전에 철저히 점검된다.
The aircraft is inspected thoroughly before each flight.

❼ 비상등은 전원 고장 시 작동된다.
The emergency lights are activated during power failures.

PART 1.4 3형식(완전타동사)

1) 문장분석

❶ Monitor the fuel efficiency regularly.
3V O
연료 효율성을 정기적으로 모니터링한다.

❷ Adjust the flight control surfaces precisely.
3V O
비행조종면을 정확하게 조절한다.

❸ The turbine delivers constant rotational energy.
S 3V O
터빈은 일정한 회전 에너지를 제공한다.

❹ Do not operate the equipment beyond its capacity.
3V O
장비를 용량 이상으로 작동하지 않는다.

❺ Stabilize the hydraulic pressure within the acceptable range.
3V O
유압을 허용 범위 내에서 안정화시킨다.

❻ The generator produces 400 Hz AC power for onboard systems.
S 3V O
발전기는 탑재 시스템을 위해 400Hz 교류 전력을 생성한다.

❼ Check that all connections are secure before starting the engine.
3V O
엔진을 시작하기 전에 모든 연결이 안전한지 확인한다.

2) 문장완성

❶ 주요 수리 후 배선이 온전한지 확인한다.
Verify that the wiring is intact after every major repair.

❷ 누설이나 손상이 있는지 연료 라인을 철저히 점검한다.
Inspect the fuel line thoroughly for any leaks or damage.

❸ 지시기는 디스플레이 패널에 현재 상태를 보여준다.
The indicator shows the current status on the display panel.

❹ 체크리스트는 매 비행 전에 필요한 점검을 명시한다.
The checklist specifies the required checks before every flight.

❺ 엔진을 끈 직후에 뜨거운 엔진 부품을 만지지 않는다.
Do not touch the hot engine parts immediately after shutdown.

❻ 추가 문제를 방지하기 위해 손상된 부품을 즉시 교체한다.
Replace the damaged component immediately to avoid further issues.

❼ 절대적으로 필요하지 않은 한, 작동 중에 패널을 제거하지 않는다.
Do not remove the panel during operation unless absolutely necessary.

PART 1.5 4형식(수여동사)

1) 문장분석

❶ The sensor sends the system real-time data.
S 4V IO DO
센서는 시스템에 실시간 데이터를 보낸다.

❷ The tool grants the motor greater precision.
S 4V IO DO
도구는 모터에 더 큰 정밀성을 부여한다.

❸ The module gives the apparatus stable power.
S 4V IO DO
모듈은 장치에 안정적인 전력을 준다.

❹ The computer shows the engineer a clear diagram.
S 4V IO DO
컴퓨터는 엔지니어에게 선명한 다이어그램을 보여준다.

❺ The alert notifies the pilot that a main door is open.
S 4V IO DO
경보는 파일럿에게 주요 문이 열려 있다고 알린다.

❻ The manual gives the equipment increased reliability.
S 4V IO DO
매뉴얼은 장비에 증가된 신뢰성을 제공한다.

❼ The software informs the user
S 4V IO
that a new update is available.
DO
소프트웨어는 사용자에게 새로운 업데이트가 가능하다고 알린다.

2) 문장완성

❶ 경보는 운영자에게 즉각적인 경고를 준다.
The alert gives the operator an immediate warning.

❷ 시스템은 차량에 최적의 성능을 보여준다.
The system shows the vehicle optimal performance.

❸ 컨트롤러는 항공기에 향상된 안정성을 제공한다.
The controller offers the aircraft enhanced stability.

❹ 모니터는 장비에 작동 상태를 보여준다.
The monitor shows the equipment operational status.

❺ 보고서는 승무원에게 점검이 완전히 완료되었다고 알린다.
The report informs the crew that the check is fully complete.

❻ 제어 장치는 정비 팀에게 긴급 경고를 보낸다.
The control unit sends the maintenance crew an urgent alert.

❼ 시스템은 운영자에게 정기적인 정비가 필요하다고 말한다.
The system tells the operator that regular maintenance is needed.

PART 1.6 5형식(불완전타동사)

1) 문장분석

❶ Make sure that the fuel tank is fully sealed.
5V OC O
연료 탱크가 완전히 밀폐되었는지 확인한다.

❷ The procedure keeps the engine secure.
S 5V O OC
절차는 엔진을 안전하게 유지한다.

❸ Do not let the oil level drop below minimum.
5V O OC
오일 수준이 최소치 이하로 떨어지지 않도록 한다.

❹ Do not allow the hydraulic system to overheat.
5V O OC
유압 시스템이 과열되지 않도록 한다.

❺ The update lets the software
S 5V O
execute the commands.
OC
업데이트는 소프트웨어가 명령을 실행하게 한다.

❻ The update makes sure
S 5V OC
that the software is current.
O
업데이트는 소프트웨어가 최신 상태임을 확실하게 한다.

❼ The checklist asks the crew
S 5V O
to verify all safety measures.
OC
체크리스트는 승무원이 모든 안전 조치를 확인하도록 요구한다.

2) 문장완성

❶ 절차는 연료 라인을 깨끗하게 유지한다.
The procedure keeps the fuel lines clear.

❷ 압력이 필요한 수준 이하로 떨어지지 않도록 한다.
Do not let the pressure drop below the required level.

❸ 냉각 시스템이 오염되지 않도록 한다.
Do not allow the cooling system to become contaminated.

❹ 업데이트는 항법 시스템이 경로를 계산하게 한다.
The update lets the navigation system calculate the route.

❺ 업데이트는 비상등이 작동하는지 확실하게 한다.
The update makes sure that the emergency lights are functional.

❻ 항공 전자기기실이 제대로 환기되고 있는지 확인한다.
Make sure that the avionics bay is properly ventilated.

❼ 체크리스트는 정비 팀이 모든 부품을 점검하도록 요구한다.
The checklist asks the maintenance team to inspect all components.

PART 2
Structural Analysis of English Sentences

영어 문장의 구조 분석

Part 2에서는 실제 항공기 정비 실무에서 사용된 문장을 분석하는 데 중점을 둔다. 이를 통해 영어 기초 학습자들이 반드시 알아야 할 문법 지식과 어휘의 용법을 익히도록 한다. 각 단원의 끝에는 영어 영작 훈련을 통해 문장의 구조를 이해하는 능력을 키운다. 실제 항공기 정비 매뉴얼에서 인용한 문장을 사용하여 학습자들이 실전에서 바로 적용할 수 있는 독해 능력을 극대화하는 데 도움을 준다.

2.1 ATA27 Flight Control System

2.2 ATA28 Fuel System

2.3 ATA29 Hydraulic Power System

2.4 ATA32 Landing Gear System

2.5 ATA36 Pneumatic System

PART 2.1 ATA27 Flight Control System

2.1.1 개요

The flight controls keep the airplane at the necessary altitude during flight. They have movable surfaces on the wing and the empennage. These are the two types of flight control systems: primary, secondary. The primary flight control system moves the airplane about three axes, lateral, longitudinal, and vertical. The primary flight control system has these components: aileron, elevator, rudder. The secondary flight controls make the lift and handling properties of the airplane better. The secondary flight control system has these components: leading edge devices, trailing edge flaps, spoilers and speed brakes, horizontal stabilizer.

[Words and Phrases]

aileron 에일러론, 도움날개
edge 모서리
flap 플랩
lateral 측면의, 가로 방향의
rudder 러더, 방향타
vertical 수직의
altitude 고도
elevator 엘리베이터, 승강타
flight controls 비행 조종
lift 양력
speed brakes 스피드 브레이크
wing 날개
axis (사물의) 축 *axes(복수형)
empennage 후방동체
horizontal stabilizer 수평 안정판
longitudinal 전후방의, 세로 방향의
spoilers 스포일러

[Translation]

비행 조종 장치는 비행 중 비행기를 필요한 고도로 유지시킨다. 그들은 날개와 후방동체에 움직이는 면을 갖고 있다. 비행 조종 시스템에는 두 가지 유형이 있다: 주 조종, 보조 조종. 주 비행 조종 시스템은 3개의 축 주위로 움직인다. 즉, 가로축, 세로축, 수직축이다. 주 비행 조종 시스템에는 다음과 같은 구성품이 있다: 에일러론, 엘리베이터, 러더. 보조 비행 조종 장치는 비행기의 양력과 조종 특성을 향상시킨다. 보조 비행 조종 시스템에는 다음과 같은 구성품이 있다: 앞쪽 모서리 장치, 뒤쪽 플랩, 스포일러와 스피드 브레이크, 수평 안정판.

2.1.2 문장분석

01

The flight controls keep[분석1)] the airplane at the necessary altitude during[분석2)] flight.
S / 5V / O / OC

분석 1 keep

동사 'keep'은 다양한 의미로 사용될 수 있다. 단지 '유지하다'의 의미로만 알고 있다면, 'keep'이 다양한 형식으로 사용될 수 있다는 점을 숙지하여 올바르게 해석해야 한다. 위 분석문의 'keep'은 아래 ⑤의 용법에 해당한다.

① **1V(keep) + to 명사:** (약속, 일정, 규칙)을 고수하다, 준수하다

Please keep to the safety protocols at all times.
항상 안전 프로토콜을 준수하기 바랍니다.

② **2V(keep) + SC(동명사):** 계속해서 ~을 하다

The warning light keeps flashing repeatedly.
경고등은 반복적으로 깜빡인다.

③ **5V(keep) + O + OC(현재분사):** ~를 ~하도록 유지하다

The system keeps the hydraulics functioning properly.
시스템은 유압 장치가 정상적으로 작동하도록 유지한다.

④ **5V(keep) + O + OC(과거분사):** ~를 ~된 상태로 유지하다

The system keeps the software updated automatically.
시스템은 소프트웨어를 자동으로 업데이트된 상태로 유지한다.

⑤ **5V(keep) + O + OC(전치사구):** ~를 ~인 상태로 유지하다

Keep the ring just above the jack collars during the entire lifting process.
전체 리프팅 과정 동안 링을 잭 칼라 바로 위에 유지한다.

분석 2 during, for

전치사 'during'과 'for'는 모두 '~동안'이라는 의미가 있으며, 시간의 기간을 나타내는 전치사이다. 그러나 이 둘의 명확한 차이점은 다음과 같다.

❶ **during:** 기간에 발생한 사건이나 상황의 명칭

The opposite occurs during power reductions.
전력 감소 중에는 반대 현상이 발생한다.

❷ **for:** 기간에 든 시간의 길이

Close the fan cowl panels and secure them with hand pressure for ten seconds.
팬 카울 패널을 닫고 손으로 압력을 가해 10초 동안 고정한다.

02

They have movable[분석1] surfaces on[분석2] the wing and the empennage.
S 3V O

분석 1 접미사 -able

'movable'은 동사 'move'에서 파생된 형용사이다. 접미사 '-able'은 특정 동사 뒤에 붙어 그 동사를 형용사로 바꾼다. 항공정비 현장에서 자주 사용되는 '-able' 접미사를 가진 형용사들의 사용 예는 다음과 같다.

❶ **accept 받아들이다** → acceptable(형) 허용 가능한

The levels of cabin pressure must remain within acceptable limits during flight.
비행 중 객실 압력 수준은 허용 가능한 범위 내에 있어야 한다.

❷ **access 접근하다** → accessible(형) 접근 가능한

Ensure that all emergency exits are accessible to passengers at all times.
모든 비상구가 항상 승객에게 접근 가능하도록 한다.

❸ **maintain 유지보수하다** → maintainable(형) 유지보수 가능한

All components should be maintainable without the need for special tools.
모든 부품은 특별한 도구 없이도 유지보수할 수 있어야 한다.

❹ **measure 측정하다** → measurable(형) 측정 가능한

Fuel efficiency should be measurable to ensure optimal performance.
연료 효율성은 최적의 성능을 보장하기 위해 측정 가능해야 한다.

⑤ **vary 다양하다** → variable(형) 가변적인

Engine performance can be variable depending on altitude and speed.
엔진 성능은 고도와 속도에 따라 가변적이다.

분석 2 전치사 on

전치사 'on'은 다양한 의미로 사용되며, 명사가 뒤에 붙어 전치사구로 사용된다. 위 분석문의 'on'은 아래 ③의 용법에 해당한다.

① **방향**

access panel on the left side
왼쪽에 있는 접근 패널

② **주제**

emphasis on safety procedures
안전 절차에 대한 강조

③ **접촉**

a small scratch on the engine cover
엔진 덮개에 있는 작은 흠집

④ **작동**

not on operation during maintenance
정비 중 작동하지 않는 상태

⑤ **일정**

scheduled maintenance check on Friday
금요일에 예정된 정기 점검

⑥ **위치**

organized tools on the maintenance cart
정비 카트 위에 정리된 도구들

03

3. The primary flight control system(S) moves[분석1](3V) the airplane(O) about[분석2] three axes, lateral, longitudinal, and vertical.

분석 1 move

동사 'move'는 1형식 완전자동사와 3형식 완전타동사의 역할을 모두 수행할 수 있다. 위 분석문의 'move'는 아래 ②의 용법에 해당한다.

① **1V(이동하다)**

The landing gear moves slowly during retraction.
착륙 장치는 접히는 동안 천천히 움직인다.

The trim tab moves to adjust the aircraft's pitch.
트림 탭이 항공기의 피치를 조정하기 위해 움직인다.

The aileron moves smoothly when the control wheel is turned.
조종 휠을 돌릴 때 에일러론이 부드럽게 움직인다.

❷ 3V(움직이다)

The valve actuator piston moves up the shaft.
밸브 액추에이터 피스톤이 샤프트 위로 이동한다.

Hydraulic actuators or electric motors move the surfaces.
유압 액츄에이터 또는 전기 모터가 표면을 움직인다.

Increased pressure from the EHSV moves the actuator to open the valve.
EHSV의 증가된 압력이 밸브를 열기 위해 액추에이터를 움직인다.

분석 2 전치사 about

전치사 'about'의 유형은 크게 네 가지로 분류할 수 있으며, 위 분석문의 'about'은 아래 ❶의 용법에 해당한다.

❶ about(주변에) + 주변 사물, 장소

Safety cones were placed about the aircraft's wings.
안전 콘은 항공기 날개 주변에 배치되었다.

❷ 감정형용사 + about(~에 대해) + 감정의 원인

The mechanics were anxious about the broken engine part.
정비사들은 고장 난 엔진 부품에 대해 불안해했다.

❸ about(약) + 시간, 수량, 정도

The inspection report will take about three hours to complete.
점검 보고서를 완료하는 데 약 3시간이 걸릴 것이다.

❹ about(~에 대한) + 주제

The training session covered information about emergency procedures.
훈련 세션에서는 비상 절차에 대한 정보를 다루었다.

2.1.3 문장완성

Answer Keys p.78

아래 한글 문장의 의미에 맞게 제시된 표현들을 적절히 조합하여 올바른 문장을 완성하시오.

❶ 그들은 날개와 후방동체에 움직이는 면을 갖고 있다.
movable surfaces / on / they / and the empennage / the wing / have

---

---

❷ 비행 조종 장치는 비행 중 비행기를 필요한 고도로 유지시킨다.
keep / the flight controls / during flight / at the necessary altitude / the airplane

---

---

❸ 보조 비행 조종 장치는 비행기의 양력과 조종 특성을 향상시킨다.
better / of the airplane / the lift / and handling properties / make / the secondary flight controls

---

---

❹ 주 비행 조종 시스템은 3개의 축 주위로 움직인다. 즉, 가로축, 세로축, 수직축이다.
moves / longitudinal, / the primary flight control system / and vertical / the airplane / lateral, / about three axes:

PART 2.2 ATA28 Fuel System

2.2.1 개요

The fuel tanks hold fuel for use by the engines and the APU. The pressure refueling system lets you add fuel to each tank. The refueling station is on the right wing. Defuelling or transferring fuel can also be accomplished at the refueling station. The main and center tanks have 2 boost pumps (fuel pumps). The boost pumps in the center tank supply fuel at a higher pressure than the pumps in the main tanks. The engines and APU use the fuel in the center tank before they use the fuel in the main tanks.

[Words and Phrases]

APU(Auxiliary Power Unit) 보조동력장치
defuel (연료를) 배출하다
fuel tanks 연료 탱크
let ~가 ~할 수 있도록 하다
on the right wing 오른쪽 날개에
refueling station 주유소, 급유 위치, 급유 스테이션
than ~보다
at a higher pressure 더 높은 압력에서
for use 사용하기 위해
hold 수용하다, 담다, 저장하다
main and center tanks 주 탱크와 중앙 탱크
pressure refueling 압력 주입, 압력식 급유 방식
supply 공급하다
transfer 이동시키다, 전달하다

[Translation]

연료 탱크는 엔진 및 APU(보조동력장치)에서 사용하기 위해 연료를 저장한다. 압력식 급유 방식을 사용하여 각 탱크에 연료를 추가할 수 있다. 급유 스테이션은 오른쪽 날개에 있다. 급유 스테이션에서는 연료를 배출하거나 이동시킬 수도 있다. 주 탱크와 중앙 탱크에는 각각 2개의 부스터 펌프(연료 펌프)가 있다. 중앙 탱크의 부스터 펌프는 주 탱크의 펌프보다 높은 압력으로 연료를 공급한다. 엔진 및 APU는 주 탱크의 연료를 사용하기 전에 중앙 탱크의 연료를 사용한다.

2.2.2 문장분석

01

The fuel tanks hold[분석1] fuel for use by the engines and the APU.
S 3V O

분석 1 hold

동사 'hold'는 여러 가지 의미로 사용될 수 있다. 단지 '잡다'의 의미 외에도 다양한 의미로 사용될 수 있다는 점을 숙지하여 올바르게 해석해야 한다. 위 분석문의 동사 'hold'는 아래 ❸의 용법에 해당한다.

❶ (사람·사물을 특정한 위치에 오게) 유지하다

Hold the flap lever steady during the system check.
시스템 점검 동안 플랩 레버를 흔들리지 않게 유지해야 한다.

❷ (손·팔 등으로) 잡고, 쥐고, 들고, 안고, 받치고 있다

Hold the wire securely while connecting the terminals.
단자를 연결하는 동안 선을 단단히 잡고 있어야 한다.

❸ (사람·사물을) 수용하다, 담다, 저장하다

This container can hold up to 200 liters of hydraulic fluid.
이 용기는 200리터의 유압액을 저장할 수 있다.

❹ (사람·사물의 무게를) 견디다, 지탱하다

Ensure that the support beams can hold the aircraft's fuselage during repairs.
수리 중에 지지 빔이 항공기의 동체를 지탱할 수 있는지 확인해야 한다.

02 The pressure refueling system lets[분석1] you add fuel to[분석2] each tank.

S / 5V / O / OC

분석 1 사역동사, 준사역동사

5형식 동사인 사역동사는 목적격보어 자리에 원형부정사를 취하며 주어가 목적어에게 어떤 일을 하도록 허락하거나 허용하는 역할을 한다. 준사역동사는 목적격보어 자리에 to부정사구를 취해 문장을 완성한다. 위 분석문의 'let'은 사역동사로서 아래 ❶의 용법에 해당한다.

❶ 5V(사역동사 let, have, make) + O + OC(동사원형)

Have the technicians conduct a thorough check.
기술자들이 철저한 점검을 시행하도록 하시오.

Make the engineers check the aircraft's hydraulic system.
우리는 엔지니어들에게 항공기의 유압 시스템을 점검하도록 해야 한다.

Please, let the mechanics inspect the engine before the flight.
비행 전에 정비사들이 엔진을 점검하도록 하시오.

❷ 5V(준사역동사 get, help) + O + OC(to부정사구)

The system update gets the software to function more efficiently.
시스템 업데이트는 소프트웨어가 더 효율적으로 작동하도록 한다.

The manual helps the technician to locate the fault.
매뉴얼은 기술자가 고장을 찾아내도록 돕는다.

분석 2 3V + O + to/with/for

3형식 완전타동사 중 일부 동사는 목적어 뒤에 전치사 'to', 'with', 'for'가 따라온다. 이러한 동사들은 목적어와 전치사구를 함께 사용하여 동작의 대상을 명확하게 나타낸다.

❶ 3V + O + to + 명사

The fuel pump adds fuel to the combustion chamber.
연료 펌프는 연소실에 연료를 추가한다.

The avionics system connects the wire to the control unit.
항공 전자 시스템은 전선을 제어 장치에 연결한다.

❷ 3V + O + with + 명사

The hydraulic system fills the actuators with hydraulic fluid.
유압 시스템은 액추에이터에 유압 유체를 채운다.

The system inspects the components with a magnifying function.
시스템은 확대 기능으로 부품을 검사한다.

❷ 3V + O + for + 명사

Measure the gap for proper alignment.
적절한 정렬을 위해 간격을 측정해야 한다.

Check the fuel system for leaks.
연료 시스템에서 누출 여부를 점검해야 한다.

03

The refueling station (S) is (2V) on the right wing (SC)[분석1].

분석 1 주격보어

주격보어는 주어를 정의하거나 상태를 나타내는 역할을 하며, 주로 다음 다섯 가지의 형태를 취할 수 있다. 위 분석문의 주격보어는 아래 ❹의 용법에 해당한다.

❶ S + 2V + SC(형용사)

The hydraulic fluid level is sufficient.
유압 유체의 레벨이 충분하다.

The inspection results are satisfactory.
점검 결과가 만족스럽다.

❷ S + 2V + SC(명사, 명사구)

The fuel pump is a critical component.
연료 펌프는 필수적인 구성품이다.

The pre-flight check is a mandatory procedure.
비행 전 점검은 필수 절차이다.

❸ S + 2V + SC(과거분사)

The safety latch is engaged.
안전 걸쇠가 걸렸다.

The emergency lights are activated.
비상등이 켜졌다.

❹ S + 2V + SC(전치사구)

The maintenance log is in the cockpit.
정비 기록은 조종석에 있다.

The tool kit is under the seat.
도구 상자는 좌석 아래에 있다.

❺ S + 2V + SC(to부정사의 명사적 용법)

The primary objective is to ensure flight safety.
주요 목표는 비행 안전을 보장하는 것이다.

The purpose of the checklist is to verify all systems.
체크리스트의 목적은 모든 시스템을 확인하는 것이다.

2.2.3 문장완성

Answer Keys p.78

아래 한글 문장의 의미에 맞게 제시된 표현들을 적절히 조합하여 올바른 문장을 완성하시오.

❶ 급유 스테이션은 오른쪽 날개에 있다.
is / the refueling station / on the right wing

____

____

❷ 압력식 급유 방식을 사용하여 각 탱크에 연료를 추가할 수 있다.
add / you / the pressure refueling system / fuel / to each tank / lets

____

____

❸ 연료 탱크는 엔진 및 APU(보조동력장치)에서 사용하기 위해 연료를 저장한다.
by the engines and the APU / fuel / hold / the fuel tanks / for use

____

____

❹ 엔진 및 APU는 주 탱크의 연료를 사용하기 전에 중앙 탱크의 연료를 사용한다.
the fuel / in the main tanks / use / in the center tank / the engines / use / before / they / and APU / the fuel

____

____

PART 2.3 ATA29 Hydraulic Power System

2.3.1 개요

There are three independent hydraulic systems that supply hydraulic power for user systems. The main and auxiliary hydraulic systems supply pressurized fluid to these airplane systems. These systems make up the hydraulic power system. The main hydraulic systems are A and B. System A has most of its components on the left side of the airplane and system B on the right side. The auxiliary hydraulic systems are the standby hydraulic system and the power transfer unit (PTU) system. The hydraulic indicating systems show these indications in the flight compartment.

[Words and Phrases]

auxiliary 보조적인
fluid 유동성 있는, 유체
indications 표시, 지시
on the left side 왼쪽에 위치한
supply 공급
compartment 구획, 격실, 칸, 모듈
independent 독립적인
make up 구성하다
pressurized 가압(압력을 가한)
transfer 이전, 전송
components 구성품
indicating 나타내는, 지시하는
most of 대부분의
standby 대기 (중인), 예비
user system 사용자 시스템

[Translation]

사용자 시스템에 유압 동력을 제공하는 세 개의 독립적인 유압 시스템이 있다. 주요 및 보조 유압 시스템은 이 항공기 시스템에 가압 유체를 공급한다. 이러한 시스템들은 유압 동력 시스템을 구성한다. 주요 유압 시스템은 A 및 B이다. 시스템 A는 비행기의 왼쪽에 대부분의 구성품을 갖추고 있으며, 시스템 B는 오른쪽에 있다. 보조 유압 시스템에는 예비용 보조 유압 시스템과 동력 전달 장치(PTU) 시스템이 있다. 유압 지시 시스템은 비행 구성품에서 이러한 표시를 보여준다.

2.3.2 문장분석

01

There are[분석1)] three independent hydraulic systems that[분석2)] supply hydraulic power for user systems.

(are: 1V / three independent hydraulic systems: S / supply: 3V / hydraulic power: O)

분석 1 도치구문

도치구문은 문장의 통상적인 어순을 바꾸어 강조하거나 문체를 다양하게 만드는 구조이다. 위 분석문의 'there are' 뒤는 주어가 도치하여 위치하고 있다. 이 주어의 수가 단수인 경우에는 there is를 사용하며, 복수일 경우는 'there are'를 사용한다.

❶ **단수**

There is no position feedback for this valve.
이 밸브에는 위치 피드백이 없다.

There is a label with the open and close instructions.
열기 및 닫기 지침이 적힌 라벨이 있다.

❷ **복수**

There are four drive shafts on each engine.
각 엔진에는 4개의 구동축이 있다.

There are three vented oil sumps on the engine.
엔진에는 배출되는 오일통이 3개 있다.

분석 2 관계대명사 that

위 분석문의 관계대명사 'that'은 선행사 'systems'를 수식하는 역할을 한다. 'that'은 관계대명사절에서 주어 또는 목적어로 사용될 수 있어 선행사와 연결하고, 관계대명사절을 이끄는 역할을 한다.

❶ **선행사(명사) + that(주격 관계대명사) + 1/2/3/4/5V**

The technician that inspects the engines is highly experienced.
엔진을 점검하는 기술자는 매우 경험이 많다.

The aircraft that undergoes regular maintenance rarely has issues.
정기적으로 정비를 받는 항공기는 문제를 거의 겪지 않는다.

❷ 선행사(명사) + that(목적격 관계대명사) + S + 3/4/5V

The tools that the mechanic uses are stored in a secure cabinet.
정비사가 사용하는 도구는 안전한 캐비닛에 보관된다.

The parts that the engineer replaced have improved the aircraft's performance.
엔지니어가 교체한 부품은 항공기의 성능을 향상시켰다.

02

The main and auxiliary hydraulic systems (S) supply (3V) pressurized[분석1)] fluid (O) to these airplane systems.

분석 1 과거분사

과거분사는 동사에서 파생된 형용사 역할을 하며, "~된"으로 해석된다. 규칙동사와 불규칙동사로 나뉘며, 규칙동사의 과거분사는 동사에 '-ed'를 붙인 형태를 취하고, 불규칙동사의 과거분사는 각각 다른 형태를 가진다.

❶ 규칙동사

The fixed components will pass all safety checks.
고정된 구성품들은 모든 안전 검사를 통과할 것이다.

The repaired engine is functioning perfectly now.
수리된 엔진은 이제 완벽하게 작동하고 있다.

The collected data showed significant wear on the landing gear.
수집된 데이터는 착륙장치에 상당한 마모가 있음을 보여주었다.

❷ 불규칙동사

The broken wing needs immediate repair.
파손된 날개는 즉각적인 수리가 필요하다.

The torn seat cover will be replaced by the maintenance crew.
찢어진 좌석 커버는 정비 팀에 의해 교체될 것이다.

The frozen hydraulic fluid will cause a delay in the flight schedule.
얼어붙은 유압 유체가 비행 일정에 지연을 초래시킬 것이다.

03 System A has most of its components on[분석1] the left side of the airplane and system B on the right side.

(System A: S / has: 3V / most of its components: O)

분석 1 방향과 위치 전치사

❶ **앞:** in front of

The maintenance crew placed the tools in front of the control panel.
정비팀이 도구들을 제어판 앞에 두었다.

❷ **뒤:** behind

The spare parts are stored behind the hangar.
예비 부품들은 격납고 뒤에 보관되어 있다.

❸ **위:** on top of, above

The emergency lights are mounted above the exit doors.
비상등이 출입문 위에 장착되어 있다.

❹ **아래:** under, below

The access panel is situated below the cockpit.
접근 패널은 조종석 아래에 위치해 있다.

❺ **안쪽:** inside, within

The avionics equipment is stored inside the electronics bay.
항공전자 장비는 전자 장비실 안에 보관되어 있다.

❻ **바깥쪽:** outside

The maintenance team performed checks outside the cabin.
정비팀이 객실 밖에서 점검을 수행했다.

❼ **오른쪽:** on the right of, to the right of

The fuel gauge is situated to the right of the main control panel.
연료 게이지가 주요 제어판 오른쪽에 위치해 있다.

❽ **가운데:** in the middle of

The new equipment is installed in the middle of the cargo bay.
새로운 장비가 화물칸 가운데에 설치되어 있다.

⑨ **왼쪽:** on the left of, to the left of

The oxygen masks are stored to the left of the overhead bins.
산소마스크는 머리 위 수납함 왼쪽에 보관되어 있다.

2.3.3 문장완성

Answer Keys p. 78

아래 한글 문장의 의미에 맞게 제시된 표현들을 적절히 조합하여 올바른 문장을 완성하시오.

❶ 사용자 시스템에 유압 동력을 제공하는 세 개의 독립적인 유압 시스템이 있다.
that supply / hydraulic power / three independent hydraulic systems / there / for user systems / are

____

____

❷ 주요 및 보조 유압 시스템은 이 항공기 시스템에 가압 유체를 공급한다.
and auxiliary hydraulic systems / pressurized fluid / supply / the main / to these airplane systems

____

____

❸ 보조 유압 시스템에는 예비용 보조 유압 시스템과 동력 전달 장치(PTU) 시스템이 있다.
and the power transfer unit (PTU) system / the standby hydraulic system / are / the auxiliary hydraulic systems

____

____

❹ 시스템 A는 비행기의 왼쪽에 대부분의 구성품을 갖추고 있으며, 시스템 B는 오른쪽에 있다.
on the right side / system B / is / of the airplane / and / system A / its components / on the left side / has / most of

____

____

PART 2.4 ATA32 Landing Gear System

2.4.1 개요

The landing gear provides support for the airplane static and ground maneuvering conditions. The landing gear also reacts to airplane load forces that are generated during airplane movement. The landing gear's extension and retraction systems extend and retract the landing gear. The nose wheel steering system supplies the ground directional control of the airplane. The tail skid system protects the lower aft fuselage if the airplane rotates too much during takeoff and landing.

[Words and Phrases]

aft 비행기 기체 후미 부분(after)
directional control 방향 제어
during 동안
extension and retraction 펼쳐짐과 접힘
fuselage 동체
generate 생성하다
landing gear 착륙장치
load forces 하중 힘
maneuvering 조종
nose wheel 전륜, 앞바퀴
provide 제공하다
react to 반응하다
rotate 회전하다
static 정적인
steering 조향
support 지지
system 시스템, 장치
tail skid 테일 스키드
takeoff 이륙

[Translation]

착륙장치는 비행기가 정지한 상태와 지상 이동 조건에서 비행기를 지지한다. 또한 착륙장치는 비행기 이동 중에 발생하는 항공기 무게를 지지한다. 착륙장치의 펼쳐짐과 접힘 장치는 착륙장치를 펼치고 접는 역할을 한다. 또한 앞바퀴 조향 시스템은 비행기의 지상 방향 제어를 제공한다. 이륙 및 착륙 중 비행기가 과도하게 회전할 경우, 꼬리 스키드 시스템은 후방 동체를 보호한다.

2.4.2 문장분석

01 The landing gear provides support for the airplane static and[분석1)] ground maneuvering conditions.

(The landing gear: S / provides: 3V / support: O)

분석 1 등위접속사

등위접속사는 두 개 이상의 대등한 절, 구, 또는 단어를 연결한다. 주요 등위접속사에는 'and, but, or, nor, for, so, yet'가 있다. 위 분석문의 'and'는 아래 ❶의 용법에 해당한다.

❶ **and**: 그리고, ~와

Vary the frequency and record the results.
주파수를 변동시키고 결과를 기록해야 한다.

❷ **but**: 그러나, ~이지만

It is free to rotate, but cannot translate.
그것은 회전은 자유로우나 이동은 불가능하다.

❸ **or**: 또는, ~이거나

Damage to translating cowl or strut may result.
이동식 카울이나 스트럿이 손상될 수 있다.

❹ **nor**: ~가 아니며, ~도 아니다

The fuel pump does not maintain pressure nor ensure proper fuel flow.
연료 펌프는 압력을 유지하지 않으며, 적절한 연료 흐름도 보장하지 않는다.

❺ **for**: ~이기 때문에

Check the turbine blades for wear, for this affects engine performance.
터빈 블레이드의 마모를 확인해야 한다. 이는 엔진 성능에 영향을 주기 때문이다.

❻ **so**: ~해서

This square drive is 0.20-inch, so a special tool must be made to fit the hole.
이 사각형 드라이브는 0.20인치이므로, 특수공구가 구멍에 맞게 제작되어야 한다.

❼ **yet**: ~이지만

The fuel pump is functioning correctly, yet the pressure remains low.
연료 펌프는 정상적으로 작동하고 있지만 압력이 여전히 낮다.

02

The landing gear's extension[분석1] and retraction systems extend and retract the landing gear.

(S: The landing gear's extension and retraction systems / V: extend and retract / O: the landing gear)

S 3V O

분석 1 접미사 -sion, -tion

'-sion' 및 '-tion' 접미사를 취하는 명사는 동사에서 파생된 경우가 많다. 따라서 동사의 의미를 이해하면 해당 명사의 의미도 쉽게 파악할 수 있다.

❶ 접미사 -sion

Inspect the expansion tank for any leaks.
팽창 탱크에 누수가 있는지 점검해야 한다.

Monitor the fuel injection system for any malfunctions.
연료 분사 시스템의 모든 오작동을 모니터링 해야 한다.

Inspect the connection to ensure there is no corrosion.
연결 부위에 부식이 없는지 확인하기 위해 검사한다.

Check the compression of the hydraulic lines regularly.
유압 라인의 압축 상태를 정기적으로 점검해야 한다.

Ensure the compression ratio is within the specified range.
압축비가 지정된 범위 내에 있는지 확인해야 한다.

❷ 접미사 -tion

Inspect the combustion chamber for any signs of damage.
연소실에 손상 징후가 있는지 검사해야 한다.

Verify the circulation of coolant to avoid engine overheating.
엔진 과열을 방지하기 위해 냉각수 순환을 확인해야 한다.

Verify the operation of the hydraulic pump for proper function.
유압 펌프가 제대로 작동하는지 확인해야 한다.

Check the insulation on all electrical wires to prevent short circuits.
단락을 방지하기 위해 모든 전선의 절연 상태를 점검해야 한다.

Monitor the fuel filtration system to ensure it is free from contaminants.
연료 여과 시스템이 오염되지 않았는지 모니터링 해야 한다.

03 The tail skid system (S) protects (3V) the lower aft fuselage (O) if[분석1] the airplane (S) rotates (1V) too much during takeoff and landing.

분석1 종속접속사

종속접속사는 두 절을 연결하여 하나의 절이 다른 절에 종속되게 한다. 종속절에는 명사절, 형용사절, 부사절이 있다. 위 분석문의 종속접속사는 부사절을 연결하며, 대표적인 부사절 종속접속사로는 if, although, as, because, before, since, unless, until, when, while 등이 있다.

❶ **if**: 만약

If the engine overheats, shut it down immediately.
만약 엔진이 과열되면, 즉시 꺼야 한다.

❷ **although**: 비록 ~일지라도

Although the circuit breaker was reset, the electrical system still shows a fault.
비록 회로 차단기를 리셋했지만, 전기 시스템에 여전히 결함이 있다.

❸ **as**: ~한 대로

Tighten the bolts as the manual instructs.
매뉴얼이 지시하는 대로 볼트를 조여야 한다.

❹ **because**: 왜냐하면

The repair took longer because the part was hard to find.
그 부품을 찾기 어려워서 수리가 더 오래 걸렸다.

❺ **before**: ~전에

Verify the fuel levels before you start the pre-flight inspection.
사전 점검을 시작하기 전에 연료 수준을 확인해야 한다.

❻ **since**: ~이후부터

The aircraft has been parked since it returned from the last flight.
그 항공기는 마지막 비행에서 돌아온 이후부터 지상에 주기되어 있다.

❼ **unless**: 만약 ~하지 않는 한

Do not proceed unless all checks are complete.
모든 점검이 완료되지 않는 한 진행하지 않아야 한다.

⑧ **until:** ~할 때까지

Keep monitoring the system until it stabilizes.
시스템이 안정될 때까지 모니터링을 계속해야 한다.

⑨ **when:** ~할 때

Replace the filter when it is clogged.
필터가 막힐 때 교체해야 한다.

⑩ **while:** ~하는 동안에

Perform the inspection while the aircraft is on the ground.
항공기가 지상에 있는 동안 검사를 수행해야 한다.

2.4.3 문장완성

Answer Keys p. 79

아래 한글 문장의 의미에 맞게 제시된 표현들을 적절히 조합하여 올바른 문장을 완성하시오.

❶ 착륙장치의 펼쳐짐과 접힘 장치는 착륙장치를 펼치고 접는 역할을 한다.
extend / and / the landing gear's extension / and retraction systems / retract / the landing gear

____

____

❷ 또한 착륙장치는 비행기 이동 중에 발생하는 항공기 무게를 지지한다.
to the load forces / generated / also / during the airplane movement / that are / reacts / the landing gear

____

____

❸ 착륙장치는 비행기가 정지한 상태와 지상 이동 조건에서 비행기를 지지한다.
for / and / the landing gear / the airplane static / provides / support / ground maneuvering conditions

____

____

❹ 이륙 및 착륙 중 비행기가 과도하게 회전할 경우, 테일 스키드 시스템은 후방 동체를 보호한다.
too much / rotates / the tail skid system / if / and landing, / during takeoff / the lower aft fuselage / the airplane / protects

____

____

ATA36 Pneumatic System

2.5.1 개요

The pneumatic system supplies compressed air to the airplane user systems. The pneumatic manifold collects the compressed air from various sources and supplies it to the user systems. Pneumatic system controls and indications are on the P5-10 air conditioning panel. The indications and controls use 28V DC 115V AC. The air in the pneumatic system is hot and under high pressure. Make sure you depressurize the pneumatic system before you work on it.

[Words and Phrases]

collect 수집하다
conditioning panel 공기조화 패널, 공조패널
make sure 반드시 ~하도록 하다, ~임을 확인하다
pneumatic system 공압 시스템
supply 공급하다
various 다양한
compressed 압축된
controls and indications 제어 및 지시 장치
manifold 매니폴드, (내연 기관의) 다기관
sources 원천
under high pressure 고압 상태의

[Translation]

공압 시스템은 항공기 시스템에 압축공기를 공급한다. 공압 매니폴드는 여러 다양한 원천에서 압축 공기를 수집하여 사용자 시스템에 공급한다. 공압 시스템의 제어 및 지시 장치는 P5-10 공기조화 패널에 있다. 지시 및 제어는 28V DC 115V AC를 사용한다. 공압 시스템 내의 공기는 뜨거우며 고압인 상태이다. 작업하기 전에 반드시 공압 시스템을 감압시켜야 한다.

2.5.2 문장분석

01 The pneumatic system supplies compressed[분석1/2] air to the airplane user systems.
S 3V O

분석 1 과거분사

과거분사(pp)에는 규칙동사와 불규칙동사 두 가지 형태가 있다. 규칙동사의 과거분사는 '-ed'로 끝나고, 불규칙동사의 과거분사는 각각 다르다. 과거분사는 주로 완료 시제, 수동태, 형용사 세 가지 역할을 한다. 위 분석문의 과거분사 'compressed'는 ❸의 용법에 해당한다.

❶ **완료 시제:** have/has/had + pp

The landing gear has been inspected.
랜딩기어가 검사되었다.

The engine overhaul has been completed.
엔진 오버홀이 완료되었다.

❷ **수동태:** be + pp

The fuel tanks are inspected regularly.
연료 탱크는 정기적으로 검사된다.

The damaged parts were replaced.
손상된 부품들은 교체되었다.

❸ **형용사:** pp

The frozen fuel lines caused a delay.
얼어붙은 연료 라인이 지연을 초래했다.

The compressed air is used for starting the engines.
압축된 공기는 엔진 시동에 사용된다.

분석 2 현재분사

현재분사는 동사의 '-ing' 형태로, 동사 진행형과 형용사의 역할을 하며 명사를 수식하거나 보어로 사용된다.

❶ 진행 시제: be + V-ing

The engine is running smoothly.
엔진이 원활하게 작동하고 있다.

The system is monitoring the fuel levels.
시스템이 연료 수준을 모니터링하고 있다.

❷ 형용사: V-ing

The running engine needs to be checked.
작동 중인 엔진을 점검할 필요가 있다.

The blinking light indicates a warning.
깜박이는 불빛이 경고를 나타낸다.

02 Pneumatic system controls and indications (S) are (2V) on[분석1] the P5-10 air conditioning panel (SC).

분석 1 in, at, on

전치사 'in', 'at', 'on'은 모두 위치를 표현하는 전치사이다. 'in'은 넓은 공간이나 내부를 나타낼 때 사용하고, 'at'은 위치나 수치의 특정한 지점을 가리킬 때 사용하며, 'on'은 표면을 나타낼 때 사용한다.

❶ in

Relieve the tension in the cable.
케이블의 장력을 완화해야 한다.

Load the spring in the cartridge.
카트리지에 스프링을 장착해야 한다.

This action must not be done in the hangar.
격납고에서는 이 행동을 해서는 안 된다.

❷ at

Cure the sealant at a uniform temperature.
일정한 온도에서 실런트를 경화시켜야 한다.

To arrive at the correct value, set the switch to position 2.
올바른 값을 얻으려면 스위치를 위치 2로 설정해야 한다.

Ensure that the keyway is at the top and in alignment with the peg.
키 홈이 상단에 있고 페그와 정렬되어 있는지 확인해야 한다.

❸ on

It is retained on the shaft by a nut.
너트로 샤프트에 고정된다.

Do not put excess weight on the trolley.
트롤리에 초과 중량을 싣지 않아야 한다.

Brackets are mounted on engine flanges.
브래킷은 엔진 플랜지에 장착된다.

03 The air (S) in the pneumatic system[분석1] is (2V) hot and under high pressure (SC).

분석 1 전치사구

전치사구는 수식어구의 일종으로 문장의 시작, 주어와 동사 사이, 목적어 뒤, 그리고 문장의 끝에 사용될 수 있다. 위 분석문의 전치사구 'in the pneumatic system'은 아래 ❷의 용법에 해당한다.

❶ 문장의 시작

On the actuator arm, a loose bolt was discovered.
액추에이터 암에서 느슨한 볼트가 발견되었다.

On the pneumatic valve, the technician noticed a leak.
공압 밸브에서 기술자가 누설을 발견했다.

❷ 주어와 동사 사이

The actuator, at the pneumatic station, responded slowly.
액추에이터가 공압 스테이션에서 느리게 반응했다.

The indicator, at the control panel, showed a high-pressure warning.
제어판에서 지시등이 고압 경고를 나타냈다.

❸ 목적어 뒤

The system monitored the air pressure in the supply lines.
시스템이 공급 라인의 공기 압력을 모니터링 했다.

The device recorded the airflow in the main duct.
장치가 주 덕트의 공기 흐름을 기록했다.

❹ 문장의 끝

The fault was detected in the pneumatic actuator.
결함이 공압 액추에이터에서 감지되었다.

The component was replaced in the air supply line.
구성품이 공기 공급 라인에서 교체되었다.

2.5.3 문장완성

Answer Keys p. 79

아래 한글 문장의 의미에 맞게 제시된 표현들을 적절히 조합하여 올바른 문장을 완성하시오.

❶ 고압 시스템 내의 공기는 고온이며, 고압인 상태이다.
hot / the air / is / in the pneumatic system / under high pressure / and

____

____

❷ 당신이 작업하기 전에 반드시 공압 시스템을 감압시켜야 한다.
you / work / depressurize / the pneumatic system / before / you / on it / make sure

____

____

❸ 공압 시스템은 항공기 시스템에 압축공기를 공급한다.
to the airplane user systems / supplies / the pneumatic system / compressed air

____

____

❹ 기압 시스템의 제어 및 표시 장치는 P5-10 공기조화 패널에 있다.
on the P5-10 air conditioning panel / and / are / indications / the pneumatic system controls

____

____

Answer Keys

PART 2.1 ATA27 Flight Control System

2.1.3 문장완성

❶ 그들은 날개와 후방동체에 움직이는 면을 갖고 있다.
They have movable surfaces on the wings and the empennage.

❷ 비행 조종 장치는 비행 중 비행기를 필요한 고도로 유지시킨다.
The flight controls keep the airplane at the necessary altitude during flight.

❸ 보조 비행 조종 장치는 비행기의 양력과 조종 특성을 향상시킨다.
The secondary flight controls make the lift and handling properties of the airplane better.

❹ 주 비행 조종 시스템은 3개의 축 주위로 움직인다. 즉, 가로축, 세로축, 수직축이다.
The primary flight control system moves the airplane about three axes: longitudinal, lateral, and vertical.

PART 2.2 ATA28 Fuel System

2.2.3 문장완성

❶ 급유 스테이션은 오른쪽 날개에 있다.
The refueling station is on the right wing.

❷ 압력식 급유 방식을 사용하여 각 탱크에 연료를 추가할 수 있다.
The pressure refueling system lets you add fuel to each tank.

❸ 연료 탱크는 엔진 및 APU(보조동력장치)에서 사용하기 위해 연료를 저장한다.
The fuel tanks hold fuel for use by the engines and the APU.

❹ 엔진 및 APU는 주 탱크의 연료를 사용하기 전에 중앙 탱크의 연료를 사용한다.
The engines and APU use the fuel in the center tank before they use the fuel in the main tanks.

PART 2.3 ATA29 Hydraulic Power System

2.3.3 문장완성

❶ 사용자 시스템에 유압 동력을 제공하는 세 개의 독립적인 유압 시스템이 있다.
There are three independent hydraulic systems that supply hydraulic power for user systems.

❷ 주요 및 보조 유압 시스템은 이 항공기 시스템에 가압 유체를 공급한다.
The main and auxiliary hydraulic systems supply pressurized fluid to these airplane systems.

❸ 보조 유압 시스템에는 예비용 보조 유압 시스템과 동력 전달 장치(PTU) 시스템이 있다.
The auxiliary hydraulic systems are the standby hydraulic system and the power transfer unit (PTU) system.

❹ 시스템 A는 비행기의 왼쪽에 대부분의 구성품을 갖추고 있으며, 시스템 B는 오른쪽에 있다.
System A has most of its components on the left side of the airplane, and system B is on the right side.

PART 2.4 ATA32 Landing Gear System

2.4.3 문장완성

❶ 착륙장치의 펼침과 접힘 장치는 착륙장치를 펼치고 접는 역할을 한다.
The landing gear's extension and retraction systems extend and retract the landing gear.

❷ 또한 착륙장치는 비행기 이동 중에 발생하는 항공기 무게를 지지한다.
The landing gear also reacts to the load forces that are generated during the airplane movement.

❸ 착륙장치는 비행기가 정지한 상태와 지상 이동 조건에서 비행기를 지지한다.
The landing gear provides support for the airplane in static and ground maneuvering conditions.

❹ 이륙 및 착륙 중 비행기가 과도하게 회전할 경우, 테일 스키드 시스템은 후방 동체를 보호한다.
The tail skid system protects the lower aft fuselage if the airplane rotates too much during takeoff and landing.

PART 2.5 ATA36 Pneumatic System

2.5.3 문장완성

❶ 공압 시스템 내의 공기는 고온, 고압인 상태이다.
The air in the pneumatic system is hot and under high pressure.

❷ 당신이 작업하기 전에 반드시 공압 시스템을 감압시켜야 한다.
Make sure you depressurize the pneumatic system before you work on it.

❸ 공압 시스템은 항공기 시스템에 압축공기를 공급한다.
The pneumatic system supplies compressed air to the airplane user systems.

❹ 기압 시스템의 제어 및 표시 장치는 P5-10 공기 조화 패널에 있다.
The pneumatic system controls and indications are on the P5-10 air conditioning panel.

PART 3
Technical English of Aircraft Maintenance Documents

항공기 정비도서의 기술 영어

PART 3에서는 영어로 작성된 항공기 기술 도서를 이해하기 위해, 항공기 주요 부품의 기능과 작동 원리를 전문적인 기술 영어로 설명한다. 항공기 기체, 엔진, 전기전자, 통신과 항법 장치로 구분하여 항공기 시스템의 기본개념을 설명한다. 간단한 정비 절차와 이해를 돕기 위한 팁을 추가하였다.

PART 3.1 Aircraft Structure

3.1.1 Structure

1) General Description

1.1 Fuselage

The fuselage is the cental of the body which provides the structure connection for the wings and tail assembly. It is a semi-monocoque construction utilizing aluminum skins, circumferential frames and longitudinal hat section stringers.

The nose radome is an aerodynamic fairing on the front of the fuselage. Most of the material in the radome is fiberglass. The radome area has navigation and weather radar antennas.

Words & Phrase			
	aerodynamic 공기 역학의	**circumferential** (원형 둘레를 따른) 주변의	
	fiberglass 유리섬유	**navigation** 항법	**utilize** 활용하다

Translation

비행기 동체는 비행기 중심부에서 날개와 꼬리 부분을 연결하는 역할을 한다. 알루미늄을 사용하여 기체 표면, 주변부 프레임, 동체 길이 방향의 스트링거로 구성된 세미 모노코크 구조이다.

노즈 레이돔은 동체 앞부분의 공기역학적 덮개이다. 대부분 유리섬유를 사용하여 제작되며, 항법과 기상에 필요한 레이더가 장착되어있다.

These dimensions give locations on the fuselage. The scale for each dimension is inches. (1) Station line (2) Body buttock line (3) Water line.

The body station line (STA) is a horizontal dimension. It starts at station line zero. You measure the body station line from a vertical reference plane that is forward of the airplane.

The body buttock line (BL) is a lateral dimension. You measure the buttock line to the left or right of the airplane center line.

The water line (WL) is a height dimension. You measure the water line from a horizontal reference plane below the airplane.

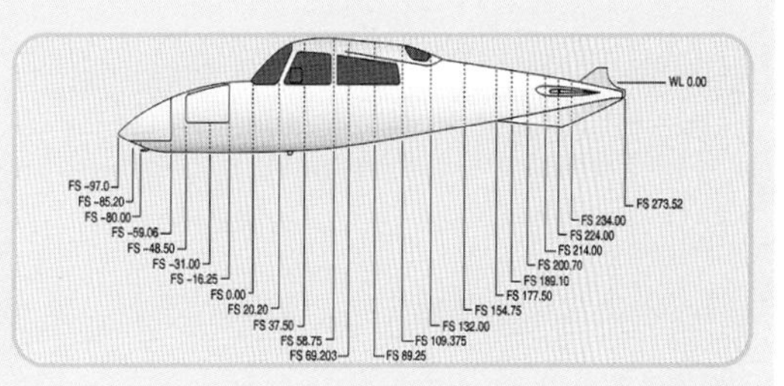

[그림 3-1] Station Line

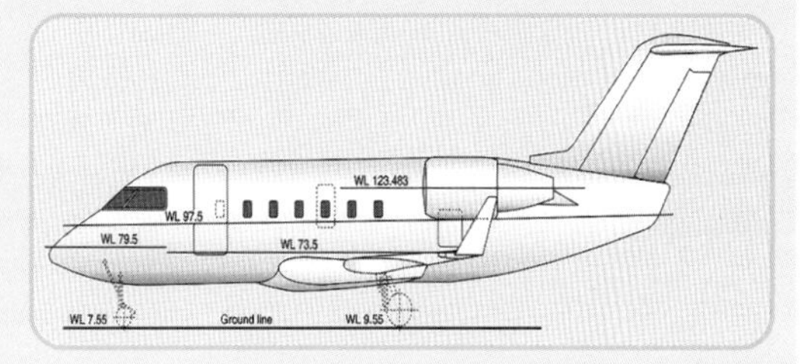

[그림 3-2] Water Line

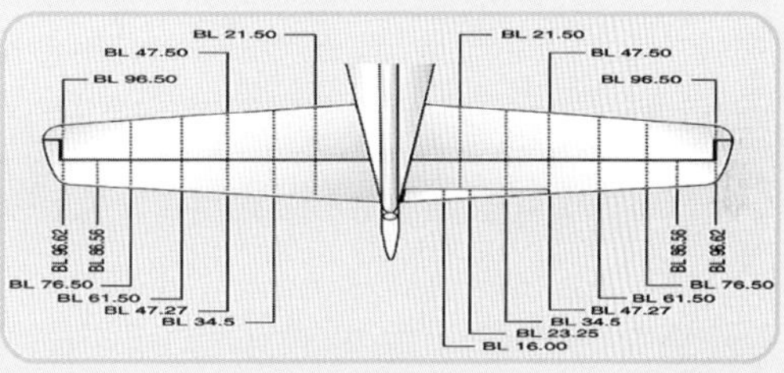

[그림 3-3] Buttock Line

Words & Phrase			
	dimension 크기, 규모, 차원	**horizontal** 수평의	**scale** 치수
	start at ~에서 시작되어	**vertical** 수직의	

Translation

다음의 치수는 동체의 위치를 알 수 있게 해준다. 각 치수의 단위는 인치를 사용한다.

(1) 스테이션 라인 (2) 바디 버턱 라인 (3) 워터 라인

바디 스테이션 라인은 수평적 차원이다. 0을 기준으로 시작하고 비행기 전방의 가상 지점에 수직면을 세워서 측정한다.

바디 버턱 라인은 측면 치수이다. 동체의 중심선을 기준으로 왼쪽 또는 오른쪽에 버턱 라인을 측정한다.

워터 라인은 높이 차원이다. 비행기 아래에 물이 차오르는 것을 가정하여 비행기 아래 수평 기준면으로부터 워터 라인을 측정한다.

Tip

station line 비행기 맨 앞에서 가장의 지점을 기점으로 하여 동체 길이 방향으로 비행기 위치를 구분하기 위한 선
buttock line 항공기 기체의 가운데를 기점으로 왼쪽, 오른쪽 구분하기 위한 선
water line 비행기 기체에 물이 차오르는 것을 가정하여 기체 아래를 기점으로 높이를 구분하기 위한 선

1.2 Nacelles/Pylons

The nacelle is the fairings and the components that surround the engine. The nacelle gives an aerodynamically smooth surface to the strut and engine. The nacelle contains fan cowl, thrust reverser and exhaust nozzle, plug.

The pylon is the strut that attaches the engine to the wing. It has fuel, electrical, hydraulic lines.

Words & Phrase			
	attach to ~에 붙이다	**contain** ~을 포함하다	**fairing** 덮개
	strut 지지대	**surround** 둘러싸는 것, 환경	

Translation

나셀은 비행기 부품의 한 종류로서 엔진 주위를 둘러싼 덮개이다. 나셀은 엔진과 지지대 주변으로 공기역학적으로 부드러운 표면을 형성한다. 나셀은 엔진 팬 카울, 역추진 장치, 배기구의 노즐과 플러그를 포함한다.

파일론은 날개에 엔진을 장착하기 위한 지지대를 말하며, 내부에 연료, 유압, 전기공급 라인들이 있다.

[그림 3-4] Nacelle and Pylon

1.3 Door

The doors are movable units that give access to the airplane compartments. we can open and close entry, galley service, and cargo doors in winds up to 40knots without structural damage. You can let these doors stay latched open in winds up to 65knots without structural damage. If a door is open for a long time, a protective cover should be put over the door frame. This prevents bad weather damage to the airplane.

When an entry and galley service doors are open and not used, a safety strap must be attached in the doorway.

Words & Phrase		
	compartment 분리된 공간, 칸막이	**galley** (비행기) 조리실
	give access 접근을 허용하다	**in wind** 바람을 타고
	latched 걸쇠를 걸다	**movable** 움직일 수 있는

Translation

비행기의 도어는 기체 내부로 출입할 수 있는 움직이는 장치이다. 출입구, 갤리 서비스, 화물용 도어는 외부 바람의 속도가 40노트일 때까지 구조적인 손상 없이 여닫을 수 있다. 비행기 도어는 최대 65노트까지 구조적 손상없이 열린 상태로 고정할 수 있다. 만약에 도어를 오랫동안 열어두려면, 프레임 위에 보호 덮개를 씌워 악천후로 인한 비행기의 손상을 방지해야 한다.

출입도어와 갤리 서비스 도어는 열린 채로 사용하지 않을 때 안전 스트랩을 부착한다.

Tip

knot 노트, 1시간에 1해리(1.85km)의 거리를 움직인 속력의 단위
safety strap 비행기 도어 개방 시 안전을 위해 설치하는 안전끈

[그림 3-5] Door Safety Strap

1.3.1 Entrance Door

The entry doors are on the left side of the airplane. The forward entry door is on the left side of the upper, forward fuselage. The forward entry door is the largest passenger entry door on the airplane.

There are two galley service doors on the airplane, forward and aft. The doors give access to the airplane passenger cabin. The doors are on the right side of the upper fuselage, across from the passenger entry doors.

The emergency exit doors are above the wings on both sides of the airplane. Emergency exit doors supply additional exits for the passengers if there is an emergency. They have the same construction features, but are adjusted separately to fit their fuselage frames.

Words & Phrase		
	across 교차하여	**additional** 부가적인
	aft 뒤쪽	**feature** 특징
	forward 앞으로	**passenger** 승객

Translation

승객의 주 출입도어는 비행기의 왼편에 위치한다. 맨 앞쪽의 출입도어는 동체의 전방에서 위쪽에 있으며, 비행기 출입도어 중에서 가장 크다.

갤리의 서비스 도어는 앞쪽과 뒤쪽으로 두 개가 있다. 출입도어를 통해 비행기 기내로 들어갈 수 있으며, 승객 출입도어의 건너편, 상부 동체 오른쪽에 위치한다.

비상탈출 도어는 비행기 날개 위쪽에 위치하며 양옆으로 나 있다. 비상 상황에서 승객이 탈출할 수 있도록 통로를 제공한다. 같은 특징을 갖고 있지만, 동체의 프레임에 알맞게 각각 조절된다.

Tip

galley service door
기내의 위치한 갤리(주방)에 필요한 용품을 공급하기 위한 서비스 카트가 드나들며, 기내에 필요한 담요, 베개 등의 서비스 용품을 공급하거나 반출하는 목적으로 주로 사용된다.

1.3.2 Service Door

The cargo doors are on the right side of the airplane. These are the two cargo doors (1) Forward cargo door, (2) Aft cargo door.

The cargo compartment doors are on the right side of the fuselage, on the lower lobe, forward and aft of the wing. The doors are similar in shape, design, and operation, but they are slightly different in size. You open and close the cargo door manually. The door has hinges on their upper edge, and swings open in a upward and inward motion.

The miscellaneous exterior service doors give access to areas with components that require regular servicing. The miscellaneous access doors are near the systems they serve.

(1) Water/Waste service door (2) Engine oil tank access door (3) APU cowl door

Words & Phrase		
	inward 안쪽으로	**lobe** 둥근 돌출부
	miscellaneous 다양한 종류의, 이것저것	**slightly** 약간
	swing open 문이 빠르게 열리는	

Translation

카고 도어는 비행기의 오른편에 있다. 앞쪽에 있는 도어와 뒤쪽에 있는 도어 두 개가 있다. 카고 도어는 날개의 앞쪽과 뒤쪽으로 동체의 돌출된 부분에 있다. 도어는 형태, 설계구조, 작동방식 등이 유사하지만 크기는 약간 다르다. 카고 도어는 수동으로 열거나 닫을 수 있다. 도어의 위쪽에는 힌지가 있어 위로 열리면서 안쪽으로 이동할 수 있는 구조이다.

비행기의 외부에 있는 다양한 종류의 도어는 시스템과 관련된 부위에 액세스 도어가 있다.

1) 물/오물 서비스 도어 2) 엔진 오일탱크 접근 도어 3) 보조동력장치 도어

[그림 3-6] Cargo Door

1.3.3 Crew Door

The crew door separates the flight compartment from the passenger compartment. The crew door is at the entrance to the flight compartment. The crew door is not a pressure door. It swings open into the passenger compartment.

The door has a mechanical and an electric lock. Access is provided by using a Keypad Access System consisting of a numeric keypad outside the flight compartment area and a chime module and electric strike that is not accessible from outside the flight compartment.

Words & Phrase			
	chime 종소리	**consist of** ~으로 되어있다	**crew** 승무원 (조종사)
	entrance 입구	**indoor** 실내의	**numeric** 수 (숫자)

Translation

크루 도어는 기내에서 조종실과 승객의 탑승 공간을 분리한다. 크루 도어는 조종실 입구에 위치한다. 여압을 담당하는 도어가 아니며, 객실을 향해 열린다.

도어는 기계식과 전기식 잠금장치가 있다. 조종실 밖에 있는 숫자 패드와 차임벨 모듈, 조종실 밖에서는 접근할 수 없는 전기식 타격 장치로 구성된 키패드 접근 시스템을 사용한다.

Tip

electric strike
도어의 잠금 방식 중에 하나로, 전기 신호로 작동되며 내부의 솔레노이드 자화에 의해 래치가 작동하는 방식

1.3.4 Warning System

The door warning system gives the crew a visual indication when a door is not secure. A door warning system shows the crew that pressure bearing doors are closed and properly latched before flight. The door warning system gives the flight crew an indication when the mid cabin exit door is not closed, latched, and locked.

Pressure doors have silicon rubber seals. The seals do these things:

(1) Seal air and light leaks

(2) Act as acoustic and thermal barriers

(3) Supply aerodynamic smoothness.

Words & Phrase

acoustic 청각의	**barrier** 장애물, 장벽	**indication** 지시
mid 중앙의	**properly** 제대로, 적절히	**rubber** 고무
silicon 실리콘	**smoothness** 부드러움, 매끄러움	
thermal 열의, 보온의	**visual** 시각의	**warning** 경고

Translation

도어 경고시스템은 조종사에게 도어가 완전하게 닫히지 않았을 때 시각적인 표시를 제공한다. 도어 경고시스템은 승무원에게 압력 베어링 도어가 닫히고 비행 전에 적절하게 잠금장치로 고정되어 있음을 보여준다. 도어 경고시스템은 중간 객실의 출입 도어가 닫히지 않거나 잠금장치가 고정되어 있지 않을 때 승무원에게 표시를 제공한다.

압력이 작용하는 도어에는 실리콘 고무 소재의 실이 있다. 실은 공기와 가벼운 누설을 막아주고, 소음이나 열을 차단하는 역할을 하며, 공기역학적인 부드러운 표면을 제공한다.

1.4 Window

All the windows can hold cabin pressurization loads and have fail-safe properties. These are the types of windows on the airplane:

(1) Flight compartment windows (fixed and sliding)

(2) Passenger compartment windows

(3) Emergency exit windows

(4) Door-mounted windows.

Words & Phrase

fixed 확고한, 고정된 **hold** 유지하다 **load** 부하, 하중
property 특성

Translation

모든 창문은 기내의 압력 하중을 버틸 수 있고 페일 세이프 구조적 특성이 있다. 비행기 창문의 종류로는 1) 조종실 창문, 2) 승객용 창문, 3) 비상탈출 창문, 4) 도어에 장착된 창문이 있다.

Tip

fail-safe
비행기 설계의 개념으로서, 같은 기능을 하는 장치가 여러 개 있어 한 개의 장치가 손상되어도 심각한 손상을 방지할 수 있음

[그림 3-7] Flight Compartment Window

[그림 3-8] Passenger Compartment Window

1.4.1 Flight Compartment Window

The left No. 1 window is the pilot's windshield. The right No. 1 window is the co-pilot's windshield. The left and right windshields are opposite assemblies and installations. The windshields install internal to the airplane. Each windshield is a laminated assembly of layers of glass and vinyl or urethane.

The No. 2 window is openable so that it can operate the right window from outside the airplane as an emergency exit.

The left No. 3 window is the pilot's side window. The right No. 3 window is the co-pilot's side window. The design of the No. 3 windows is to carry pressure loads.

Current windshields can have a temporary rain repellant hydrophobic coating applied. Some older airplanes have an engaged or disengaged liquid rain repellant system. Windshield wipers supply a sufficient quantity of rain removal without a secondary rain repellant system.

Words & Phrase

carry (물, 전기 등을) 실어 나르다, 전달하다
engage 사로잡다, 관계를 맺는다
laminate 얇은 판을 여러 장 붙여 만든 것
repellant (물 등이) 배지 않는, 반발하는
temporary 임시의
current 현재의, 통용되는, 전류
hydrophobic 소수성의 (방수)
openable 열 수 있는
sufficient 충분한
windshield 바람막이 창

Translation

왼쪽의 1번 창문은 조종사(기장)의 앞 유리고, 오른쪽의 1번 창문은 조종사(부기장)의 앞 유리다. 왼쪽과 오른쪽 창문은 모양이 반대이고 같은 구조로 되어있다. 앞 유리는 비행기 내부에서 설치한다. 각 앞 유리는 유리와 비닐 또는 우레탄 층을 적층한 조립체이다.

2번 창문은 개폐식으로 되어 있어 비행기 외부에서 창문을 비상구로 사용할 수 있다.

왼쪽 3번 창문은 조종사(기장)의 옆 창문이고, 오른쪽 3번 창문은 조종사(부기장)의 옆 창문이다. 3번 창문은 압력 하중을 전달하기 위해 설계되었다.

현재의 윈드쉴드는 일시적인 방수용액 코팅이 적용된다. 일부 오래된 비행기들은 액체가 공급되는 방수 시스템과 결합하거나 분리된 시스템이 적용되며, 윈드쉴드의 와이퍼는 2차 방수용액 공급 시스템과 관계없이 상당한 양의 빗물을 제거할 수 있다.

1.4.2 Passenger Compartment Window

The passenger compartment windows have these components;

1) Outer pane 2) Middle pane 3) Inner pane.

The outer and middle panes are structural. The outer pane is made of stretched acrylic plastic. It is rectangular with rounded corners and a beveled outer edge to fit with the window frame. The shape of the pane is curved to align with the fuselage contour.

The middle pane gives the structural fail-safe function. It can hold 1.5 times the normal pressure load. The middle pane is made of cast acrylic and has a shape like the outer pane, but with an edge not beveled. The middle pane is contained in the window seal. A small breather hole is near the bottom of the middle pane.

The inner pane is not structural. It connects to the cabin sidewall panel. The inner pane is made from polycarbonate.

Words & Phrase

acrylic 아크릴로 만든
breather hole 작은 구멍 (창문 습기 방지용)
contour 윤곽
stretched 신축성이 있다, 늘이다, 늘어나다
bevel 비스듬한
cast 주물
rounded (모양이) 둥근
structural 구조적인

Translation

승객용 창문은 다음의 부품으로 구성되어있다. 1) 아우터 페인 2) 미들 페인 3) 이너 페인

아우터 페인과 미들 페인은 구조적으로 되어있다. 아우터 페인은 늘어진 아크릴 플라스틱으로 되어있다. 창틀에 맞게 모서리가 둥글고 바깥쪽 가장자리가 경사져 있는 직사각형이다. 페인의 모양은 동체 윤곽에 맞게 곡선으로 되어있다.

미들 페인은 구조적인 페일 세이프 기능을 제공한다. 그것은 일반 압력 부하의 1.5배를 견딜 수 있다. 미들 페인은 주조 아크릴로 만들어졌으며 아우터 페인과 같은 모양을 가지고 있지만, 가장자리가 경사지지 않은 상태이다. 미들 페인은 창틀에 들어 있다. 미들 페인의 바닥 근처에 작은 구멍이 있다.

이너 페인은 구조역학적으로 힘을 받지 않으며, 객실의 옆면 벽의 패널과 함께 장착되어 있다. 폴리카보네이트 재료를 이용하여 제조된다.

2) Maintenance Procedure

2.1 Repair the aft fairing access panels high temperature enamel paint

a. strip the existing enamel finish and primer Paint Stripping

b. touch up surface with chemical conversion coating

c. apply one coat of primer

d. apply high temperature enamel, the cure time before flying away is 24 hours, 10 days for full cure.

Words & Phrase			
	finish 마감재	**one coat** 1회 도포	**primer** 프라이머, 밑칠 페인트
	repair 수리하다	**strip** 벗겨내다, 제거하다	**surface** 표면
	touch up 부분 수정하다		

Translation

2.1 엔진 애프터 페어링 패널의 고온용 에나멜 수리하기

a. 기존 에나멜 마감재와 프라이머 페인트를 제거한다.

b. 화학적 변환 코팅으로 표면을 부분 수정한다.

c. 프라이머를 1회 도포한다.

d. 고온에 사용하는 에나멜을 바른다. 에나멜이 굳는 시간은 비행 전까지 24시간이며, 완전히 굳기 위해서는 10일이 필요하다.

[그림 3-9] CFM56-7B Engine

[그림 3-10] Pylon After Fairing

2.2 Forward Entry Door Pressure Seal Check

a. make sure the door is safe.

b. make sure the door is closed and latched.

MAKE SURE THE GIRT BAR IS NOT ENGAGED IN THE FLOOR MOUNTED ESCAPE SLIDE BRACKETS. IF THE GIRT BAR IS ENGAGED IN THE BRACKETS, THE ESCAPE SYSTEM CAN DEPLOY WHEN YOU OPEN THE DOOR.
THIS CAN CAUSE INJURIES TO PERSONS AND DAMAGE TO EQUIPMENT.

c. make sure the girt bar is not engaged in the floor-mounted escape slide brackets.

d. make sure a work platform, is installed outboard of the door.

e. open the door.

f. do a visual inspection of the door pressure seal

g. examine the seal.

  1) look for cracks, holes, and tears.

  2) look for indications of seal deterioration.

  3) make sure the seal is installed in the seal retainer.

h. close and latch the door.

i. remove the work platform

Words & Phrase

conversion 전환	**deploy** 배치하다
deterioration 악화되다, 나빠지다	**engage** 맞물려있다, (주의, 관심을) 사로잡다
flying away 비행 전	**tear** 찢다, 구멍을 뚫다
work platform 작업대	

Translation

2.2 승객용 도어 압력 실 점검

a. 도어가 안전한지 확인한다.

b. 문이 닫혀 있고 걸쇠로 고정되어 있는지 확인한다.

WARNING 〈경고〉 거트 바가 바닥에 장착된 비상탈출 슬라이드 받힘대에 결합하지 않았는지 확인한다. 거트 바를 연결하면, 문을 열 때 비상탈출 시스템이 작동하여 슬라이드가 펼쳐질 수 있다. 이로 인해 사람이 다치거나 장비가 손상될 수 있다.

c. 거트 바가 바닥에 장착된 탈출 슬라이드 받힘대에 결합하지 않았는지 확인한다.

d. 도어 바깥쪽에 작업대가 설치되어 있는지 확인한다.

e. 문을 연다.

f. 도어 압력 실을 육안으로 검사한다.

g. 실을 확인한다.

1) 균열, 구멍, 찢어진 곳을 찾는다.

2) 실의 상태가 악화한 징후가 있는지 확인한다.

3) 실에 실 리테이너에 설치되어 있는지 확인한다.

h. 문을 닫고 잠근다.

i. 작업대를 제거한다.

Tip	**cure time** 큐어타임, 비행기 정비 시 실런트를 도포 한 후 완전히 굳는 데 걸리는 시간을 의미 **girt bar** 비행기 문과 비상탈출 미끄럼틀을 연결해주는 장치

3.1.1 Practice Quiz

Answer Keys p. 194

Q. Choose the correct word from the box to complete the sentences below.

nacelle semi-monocoque girt-bar movable fail-safe

01 It is __________ of construction utilizing aluminum skins, circumferential frames and longitudinal hat section stringers.

02 The __________ contains fan cowl, thrust reverser and exhaust nozzle, plug.

03 The doors are __________ units that give access to the airplane compartments.

04 All the windows can hold cabin pressurization loads and have __________ properties.

05 If the __________ is engaged in the brackets, the escape system can deploy when you open the door.

3.1.2 Wings

1) General Description

1.1 Main Wing

Aircraft wings are consist of main wing, stabilizer structure and flight control surface. Most of material in them is aluminum but rudder, elevator are made of graphite composite. The internal structures of the wing include spars and stringers.

Some aircraft has winglet instead of wing tip. The wing tip has anti-collision lights and forward and aft position lights. The winglets are wing tip extensions which provide several benefits to operators.

Words & Phrase

be made of ~으로 제작되다　**consist of** ~으로 구성되다　**extension** 확장, 연장, 내선 번호
instead of ~ 대신에　**operator** 운영자

Translation

비행기의 날개는 주 날개와 동체 후방의 스태빌라이저 그리고 비행조종면으로 구성된다. 날개를 제작할 때 주재료로 사용되는 것은 알루미늄이며, 러더와 엘리베이터는 복합소재를 이용하여 만들어진다. 날개의 내부 구조에는 스파와 스트링거 등이 있다.

비행기에는 윙팁 대신에 윙렛을 장착한다. 윙팁에는 비행기의 충돌을 방지하기 위한 충돌방지등이 있고, 위치를 나타내기 위한 위치지시등이 앞쪽과 뒤쪽에 있다. 윙렛은 윙팁을 확장한 것으로 비행기 운영에 장점이 있다.

Tip

consist
어떤 종류의 물건들로 모여있는 형태를 나타낼 때 사용하고, make는 어떤 사물의 재료를 의미할 때 사용한다. 비행기 부품의 세부 구성품을 설명할 때 정비사들이 두 단어를 모두 사용하는 경우가 있어 헷갈리지 않도록 주의해야 한다.
operator 항공사나 비행기를 소유하고 운영하는 주체를 뜻함
anti-collision light 충돌방지등
position light 위치지시등

1.2 Flight Control System

The flight controls keep the airplane at the necessary attitude during flight. They have movable surfaces on the wing and the empennage. These are the two types of flight control systems: primary and secondary

The primary flight control system moves the airplane about three axes: lateral, longitudinal, and vertical. The primary flight control system has these components:

1) Aileron 2) Elevator 3) Rudder

The secondary flight controls improve the lift and handling properties of the airplane. The secondary flight control system has these components:

1) Leading edge flaps and slats 2) Trailing edge flaps 3) spoilers and speedbrakes

Words & Phrase

attitude 비행기 자세
handling property 조종 성능(의역)
lift 양력, 승강기
primary 주된, 주요한
secondary 이차적인, 부수적인

Translation

비행기 조종면은 비행하는 동안 비행기 자세를 유지할 수 있는 장치이다. 움직이는 조종면은 주 날개와 꼬리날개에 있다. 1차 조종면과 2차 조종면으로 구분한다.

1차 조종면은 비행기 3가지 자세축을 기준으로 비행기를 움직이게 한다. 1차 조종면을 이루는 구성품으로는 에일러론, 엘리베이터, 러더가 있다.

2차 조종면은 비행기의 양력과 조종 성능을 높여 준다. 2차 조종면의 구성품으로는 리딩엣지 플랩과 슬랫, 트레일링 엣지 플랩, 스포일러, 스피드브레이크가 있다.

Tip

Flight Controls 비행기 main wing에서 움직이는 조종면을 일컫는 말
L/E leading edge의 약어 (LE 표기도 가능)
T/E Trailing egde의 약어 (TE 표기도 가능)

1.3 Flaps

The inboard and outboard flaps increase the wing area and the wing camber. This helps improve the airplane performance during takeoff and landing. Pilots use the flap lever during the normal operation of the TE flaps.

The LE flaps and slats increase the lift of the wing during takeoff and landing. They retract during cruise. The LE flaps are below the leading edge of the wings, between the fuselage and the engines. The LE slats are on the leading edge of the wings, outboard of the engines.

Words & Phrase

airplane performance 비행기 성능
inboard 안쪽 (동체와 가까운 부위)
outboard 바깥쪽 (동체와 먼 부위)
wing area 날개 면적
cruise 순항 (일정한 고도에서의 비행)
increase 증가하다
retract 들어가다, 집어넣다
wing camber 날개 곡률

Translation

안쪽의 플랩과 바깥쪽의 플랩은 날개의 면적과 곡률을 증가시킨다. 이는 비행기가 이 · 착륙할 때 비행기 성능에 도움이 된다. 조종사는 평상시 플랩 레버를 사용하여 트레일링 플랩을 작동한다.

리딩엣지 플랩과 슬랫은 비행기가 이착륙할 때 날개의 양력을 증가시킨다. 비행기가 순항 중일 때는 접혀있는 상태이다. (제자리에 위치함) 리딩엣지 플랩은 날개의 리딩엣지 아래에 위치하며 비행기 동체와 엔진 사이에 있다. 리딩엣지 슬랫은 엔진 바깥쪽으로 날개의 리딩엣지에 있다.

1.4 Spoiler

The spoilers help the ailerons control airplane roll about the longitudinal axis. They also supply speedbrake control to reduce lift and increase drag during landing and refused takeoff. The ground spoilers are the most outboard and the most in-board spoiler on each wing. All the other spoilers are flight spoilers.

All the spoilers move up when the airplane is on the ground, and only the flight spoilers move up when the airplane is in the air.

Words & Phrase

airplane roll 비행기의 롤링 (좌우로 움직이는 성능) **in the air** 비행 중
on the ground 지상에서 **reduce** 줄이다, 낮추다. **refuse** 거절하다
the most outboard 동체에서 가장 바깥쪽의

Translation

스포일러는 에일러론의 비행 롤링 조종을 돕는다. 또한, 양력을 줄이거나 증가한 항력을 줄이기 위해 착륙과 이륙 중단 시 스피드브레이크를 제어한다. 그라운드 스포일러는 각 날개의 가장 바깥쪽과 가장 안쪽의 위치한 스포일러를 의미한다. 다른 스포일러는 플라이트 스포일러라 부른다.

지상에서 모든 스포일러는 위로 작동하며, 비행 중일 때만 플라이트 스포일러가 작동한다.

Tip

refused takeoff
이륙 중단을 의미하며, 비행기가 활주로에서 이륙을 위해 앞으로 나아가다 갑자기 멈추는 경우 정지를 위해 speedbrake 역할을 하는 spoiler가 중요함

1.5 Pilot operation

The pilots manually operate the flight controls through cables. Aircraft's autopilot system automatically operates them. The flight control cables to give manual input to each flight control system. Most cables are under the floor and go from the flight compartment to the respective flight controls.

Words & Phrase

from A to B A에서 B로 **operate ~ through** ~을 통해서 운영되다
respective 각각의 **to give** ~을 주다 **under the floor** 바닥 아래에

Translation

조종사는 케이블로 연결된 조종면을 수동으로 조종한다. 비행기의 자동조종 장치는 조종면을 제어하여 움직인다. 조종면에 연결된 케이블은 수동 입력 신호를 각 조종면에 전달한다. 많은 케이블은 바닥 아래를 지나 조종실에서 각각의 조종면까지 연결되어있다.

2) Maintenance Procedure

2.1 Rudder trim actuator removal

a. disconnect the electrical connector from the rudder trim actuator
b. remove the nut, washer and bolt assembly to disconnect the rudder feel and centering unit from the rudder trim actuator.
c. remove the nut, washer, bushing and bolt that attach the rudder trim actuator to the airplane structure.
d. remove the rudder trim actuator from the airplane.

Words & Phrase

attach 붙이다
trim 다듬다, 손질하다
disconnect 공급을 끊다
structure 구조

Translation

2.1 러더 트림 액추에이터 분리

a. 러더 트림 액추에이터에서 전기 커넥터를 분리해라.
b. 러더 트림 액추에이터에서 필 앤 센터링 유닛을 분리하기 위해 관련된 너트, 볼트, 와셔를 제거하라.
c. 러터 트림 엑추에이터를 항공기 동체와 연결하는 너트, 와셔, 부싱, 볼트를 제거하라.
d. 비행기에서 러더 트림 액추에이터를 분리하라.

[그림 3-11] Rudder

2.2 Rudder trim actuator installation

a. make sure that these circuit breakers are open
b. make sure that you remove pressure from the rudder hydraulic systems
c. install the bolt assembly, bushing, washer and nut to connect the rudder trim actuator to the airplane structure, rudder feel and centering unit.
d. connect the electrical connector to the rudder trim actuator
e. remove the safety tags and close circuit breakers.
f. operate the rudder trim knob until you can install the rig pin.
g. if the rudder trim indicator reading is not 0.0 +/- 0.1 unit, turn the adjustment screw on the bottom of the rudder trim indicator.

Words & Phrase

adjustment 조절 **circuit breaker** 회로 차단기 **knob** 손잡이
rig 수정하다, 조작하다

Translation

2.2 러더 트림 액추에이터 장착

a. 서킷 브레이커의 오픈 상태를 확인해라.
b. 러더에 관련한 유압 시스템의 압력을 제거했는지 확인해라.
c. 비행기 동체와 러더 필 앤 센터링 유닛과 러더 트림 액추에이터를 연결하기 위해 볼트, 와셔, 부싱 등을 장착하라.
d. 전기 커넥터와 러더 트림 액추에이터를 연결하라.
e. 안전 태그를 제거하고, 서킷 브레이커를 닫는다.
f. 케이블 리그 핀을 장착 할 수 있을 때 까지 러더 트림 손잡이를 작동한다.
g. 만약, 러더 트림 지시계가 0.0 +/− 0.1 유닛을 지시하지 않으면, 러더 트림 액추에이터 아래쪽의 조절 나사를 돌려라.

2.3 Fault Isolation Manual - Rudder Does not move Correctly

(1) Initial Evaluation

a. do a check of the Rudder Travel

b. do this check of the Rudder Pedals: approximately 5 full cycles

(2) Fault Isolation Procedure

a. do a check of the Rudder Control Cables, Rudder Pedals, and Forward Quadrant Adjustment

b. do a check of the Control Rod for the Main Rudder Power Control Unit (PCU) adjustment

c. do a check of the Control Rod for the Standby Power Control Unit (PCU) adjustment

(3) Repair Confirmation

a. if the test passes, then you corrected the problem

b. if the test fails, then continue troubleshooting at the subsequent step

Words & Phrase		
	approximately 거의, 대략적으로	**make sure** 확실하게 하다
	subsequent 후속	**troubleshooting** 고장 탐구

Translation

2.3 러더가 제대로 작동하지 않을 때

(1) 초기 진단

a. 러더의 움직임을 점검하라.

b. 러더 페달을 약 5회 이상 움직여 점검하라.

(2) 고장 탐구 절차

a. 러더 컨트롤 케이블, 러더 페달, 케이블 쿼드런트 조절기 등을 확인하라.

b. 메인 러더 전력 컨트롤 유닛의 연결된 로드를 확인하라.

c. 보조 러더 전력 컨트롤 유닛의 연결된 로드를 확인하라.

(3) 수리 확인

a. 점검 결과가 성공적이면, 결함을 해소하였다.

b. 점검 결과가 안 좋으면, 세부 절차의 고장 탐구를 계속 수행해라.

3.1.2 Practice Quiz

Answer Keys p. 194

Q. Choose the correct word from the box to complete the sentences below.

attitude	are consist of	flaps	increase	troubleshooting

01 Aircraft wings ___ ______ ___ main wing, stabilizer structure and flight control surface.

02 The flight controls keep the airplane at the necessary ________ during flight.

03 The inboard and outboard ________ increase the wing area and the wing camber.

04 They also supply speedbrake control to reduce lift and ________ drag during landing and refused takeoff.

05 If the test fails, then continue ________ at the subsequent step.

3.1.3 Aircraft System

1) General Description

1.1 Air conditioning

The air conditioning system controls the interior environment of the airplane for flight crew, passengers, and equipment. The air conditioning sub-systems are distribution, pressurization, equipment cooling, heating, cooling, and temperature control.

The quantity of fresh air necessary for ventilation is more than for pressurization. The ventilation quantity is based on a fixed value for the crew and allowable leakage, and the number of passenger seats.

The distribution system has some functions.

1) Divide conditioned air into the three airplane zones

2) Reduce engine bleed requirements

3) Remove offensive air from lavatories and galleys

4) Supplies cooling air to electronic equipment

Words & Phrase			
	air-conditioning 에어컨	**distribution** 분배	**environment** 환경
	offensive 불쾌한, 역겨운	**pressurization** 가압	**ventilation** 환기

Translation

공조 시스템은 비행 승무원, 승객과 장비를 위해 비행기의 내부 환경을 제어한다. 공조 장치의 하위 시스템은 분배, 가압, 장비 냉각, 가열, 냉각, 온도 제어이다.

환기에 필요한 신선한 공기의 양은 가압되는 공기보다 많아야 한다. 환기되는 정도는 조종사와 허용된 누출 값, 승객 좌석 수를 기준으로 설정된 고정된 값을 따른다.

분배 시스템에는 몇 가지 기능이 있다.

1) 조절된 공기를 3개의 비행기 구역으로 분배

2) 엔진 블리딩 요구사항 감소

3) 화장실 및 갤리에서 불쾌한 공기를 제거

4) 전자 장비에 냉각 공기를 공급

1.1.2 Cooling and Heating

The equipment cooling system removes heat from electronic components in the flight compartment and the E/E compartment. The system uses cabin air to remove heat from equipment. Fans move the air through ducts and manifolds.

The heating system supplies warm air to areas to prevent freezing or to increase temperature for comfort. The air mixes with the passenger compartment air in the main distribution manifold. The cargo compartments receive heat from equipment cooling exhaust and passenger compartment air. warm equipment cooling exhaust air flows under the forward cargo compartment floor and along the sidewalls. The aft cargo compartment air comes from the passenger compartment through the foot-level grilles. The air goes into the sidewall area around and under the aft cargo compartment through the outflow valve.

Words & Phrase

comfort 편안, 안락
equipment 장비, 설비
exhaust 배기관
freeze 얼다, 얼리다
manifold 다기관, 수가 많은
mix 섞이다, 혼합하다
prevent 방지하다, 예방하다

Translation

장비 냉각 시스템은 조종석과 E/E 실의 전자 부품에서 열을 제거한다. 시스템은 객실 공기를 사용하여 장비에서 열을 제거한다. 팬은 덕트와 매니폴드를 통해 공기를 이동시킨다.

난방 시스템은 동파를 방지하거나 쾌적함을 위해 온도를 높여 따뜻한 공기를 기내에 공급한다. 공기는 메인 매니폴드에서 기내의 공기와 혼합된다. 화물칸은 장비를 냉각하고 배출되는 가스와 기내의 공기로부터 열을 공급받는다. 따뜻한 장비 냉각 배기 공기는 전방 화물칸 바닥 아래 및 옆벽을 따라 이동한다. 애프터 화물칸 공기는 기내에서 승객의 발 옆쪽의 그릴을 통해 공급되며, 공기는 아웃플로우 밸브를 통해 애프터 화물칸 주변과 아래의 옆쪽 벽으로 들어간다.

Tip

E/E electronic and equipment의 줄임말로서, 조종실 아래에 위치한다. 비행기 운영에 필요한 배터리와 각종 컴퓨터가 있으며 내부에서 발생하는 열기를 제거하기 위해 냉각이 중요하다.

1.1.3 Pressurization

The airplane operates at altitudes where the oxygen density is not sufficient to sustain life. The pressurization control system keeps the airplane cabin interior at a safe altitude. This protects the passengers and crew from the effects of hypoxia (oxygen starvation).

The cabin pressure control system controls the rate that at which air flows out of the cabin. The cabin pressure relief system is a fail-safe system. It protects the airplane structure from overpressure and negative pressure if the pressurization control system fails.

Words & Phrase			
	density 밀도	**hypoxia** 저산소증	**keep** ~을 유지하다
	negative pressure 부압 (대기압보다 낮은 압력)		**overpressure** 지나친 압력
	oxygen starvation 산소 부족		**sustain** 계속 시키다

Translation

비행기는 산소 밀도가 생명을 유지하기에 충분하지 않은 고도에서 작동한다. 여압 제어 시스템은 비행기 객실 내부를 안전한 고도로 유지한다. 이것은 저산소증(산소 부족)의 영향으로부터 승객들과 승무원들을 보호한다.

객실 압력 제어 시스템은 객실 밖으로 공기가 유출되는 속도를 제어한다. 객실 압력 완화 시스템은 페일 세이프 시스템이다. 이것은 가압 제어 시스템이 고장 날 때 과압 및 부압으로부터 비행기 구조물을 보호한다.

1.2 Fuel

The fuel tanks store fuel for use by the engines and the APU. It is stored in the wings of the plane and is called the main fuel tank. The pressure fueling system lets you add fuel to each tank. The fueling station is on the right wing. It also can defueling and fuel transfer at the fueling station.

Each main tank has two boost pumps (fuel pumps). The center tank also has two boost pumps. The center tank boost pumps supply fuel at a higher pressure than the pumps in the main tanks. Because of this, the fuel in the center tank is used before the fuel in the main tanks.

Words & Phrase

ambient 주위의
defuel ~에서 연료를 제거하다
vent 통풍구
boost 증가, 증가시키다
store 저장하다

Translation

연료 탱크는 엔진과 보조동력장치에서 사용할 연료를 저장한다. 비행기 날개 안에 연료가 저장되며 메인 탱크로 정의한다. 압력 연료 공급 시스템은 각 탱크에 연료를 추가할 수 있다. 연료가 공급되는 장소는 오른쪽 날개에 있다. 이곳을 통해서 연료를 회수하고 옮길 수도 있다.

각 메인 탱크에는 2개의 부스트 펌프(연료 펌프)가 있다. 센터 탱크에도 2개의 부스트 펌프가 있다. 센터 탱크 부스트 펌프는 메인 탱크의 펌프보다 높은 압력으로 연료를 공급한다. 이 때문에, 센터 탱크의 연료가 메인 탱크의 연료보다 먼저 사용된다.

1.2.1 Fueling

All tanks fill from the fueling station at the right wing. The fueling station permits automatic and manual control of the fueling shutoff valves. The fueling station receives 28v dc hot battery bus power through the refueling power control relay. The relay energizes when you open the door of the fueling station.

The fueling indication test switch is used to supply an alternative ground for the refueling power control relay. The engine fuel feed system supplies fuel to the engines from main tank 1, main tank 2, and the center tank. The boost pumps supply fuel from main tank 1 to the engine feed manifold. The boost pumps supply fuel at a minimum pressure.

Words & Phrase

alternative 대안, 대체의　**energize** 동력을 공급하다　**feed system** 공급 장치
fill 채우다　**fuel transfer** 연료 이송　**fueling station** 연료 보급 부위
hot battery bus 배터리에서 항상 전류가 공급되는 라인　**shutoff valve** 차단 밸브

Translation

우측 날개에 있는 연료 스테이션으로부터 모든 탱크가 충전된다. 연료 스테이션에는 연료 공급 차단 밸브를 자동이나 수동으로 제어할 수 있다. 연료공급 전력 제어 릴레이를 통해 28v 직류 핫 배터리 버스 전력을 공급받는다. 스테이션의 패널을 열면 릴레이에 전기가 공급된다.

연료공급 표시 테스트 스위치는 연료공급 전력 제어 릴레이의 대체 접지를 공급하기 위해 사용된다. 엔진 연료 공급계는 메인 탱크 1, 메인 탱크 2 및 센터 탱크로부터 엔진들로 연료를 공급한다. 부스트 펌프들은 메인 탱크 1로부터 엔진 공급 매니폴드로 연료를 공급한다. 부스트 펌프는 최소 압력으로 연료를 공급한다.

1.2.2 Defueling

The defuel system permits pressure defuel of each tank and suction defuel of main tank 1 and main tank 2. The defuel system also allows fuel transfer on the ground from one fuel tank to another. when you use the defuel valve, the refuel station, and the fuel control panel to transfer fuel between tanks. defuel valve connects the right engine fuel feed manifold with the defuel manifold. This permits the removal of fuel from the fuel tanks.

Words & Phrase			
	defuel ~에서 연료를 제거하다	**defueling** 배유	**permit** 허용하다
	refuel 연료를 재급유하다	**suction** 흡입, 빨아들이기	

Translation

연료 배출 시스템은 각 탱크의 압력 배출과 메인 탱크 1 및 메인 탱크 2의 흡입 배출을 허용한다. 하나의 연료 탱크에서 다른 연료 탱크로 연료를 이송할 수 있다. 연료 배출 밸브, 연료 주입 스테이션 및 연료 제어 패널을 사용하여 탱크 간에 연료를 주입할 때 연료 배출 밸브는 우측 엔진 연료공급 매니폴드와 연료 배출 매니폴드를 연결한다. 이를 통해 연료 탱크로부터 연료를 제거할 수 있다.

1.3 Hydraulic

Three independent hydraulic systems supply hydraulic power for user systems. Two independent hydraulic systems supply power to airplane systems. The standby hydraulic system supplies an alternative source of power.

The main and auxiliary hydraulic systems supply pressurized fluid to these airplane systems:

1) Both thrust reversers
2) Power transfer unit (PTU) motor
3) Landing gear extension and retraction
4) Nose wheel steering
5) Main gear brakes
6) Primary flight controls
7) Secondary flight controls

Words & Phrase			
	auxiliary 보조의	**extension** 확대	**fluid** 유체
	hydraulic 유압식의	**independent** 독립된	**standby** 예비품
	steering 조향 장치		

Translation

3개의 독립적인 유압 시스템은 사용자 시스템을 위한 유압 전력을 공급한다. 2개의 독립적인 유압 시스템은 비행기 시스템에 전력을 공급한다. 보조 유압 시스템은 대체 전력원을 공급한다. 메인 및 보조 유압 시스템은 아래의 비행기 시스템에 유압을 공급한다.

1) 양쪽의 역추진 장치
2) PTU 모터
3) 랜딩기어 작동
4) 방향 제어
5) 메인 기어 브레이크
6) 1차 조종면
7) 2차 조종면

1.3.1 Pressurization

The hydraulic reservoir pressurization system supplies filtered bleed air from the airplane's pneumatic system to the main hydraulic systems A and B reservoirs. Air pressure from the reservoir pressurization system maintains head pressure on hydraulic system A, system B, and the standby hydraulic system reservoirs. The pressurized reservoirs supply a constant flow of fluid to the hydraulic pumps.

The reservoirs can be manually pressurized using an external ground air source without the need to pressurize the pneumatic system crossover manifold. The system provides a positive supply of hydraulic fluid to the pumps, maintains normal return pressure in the hydraulic system, and also helps to prevent foaming in the reservoirs.

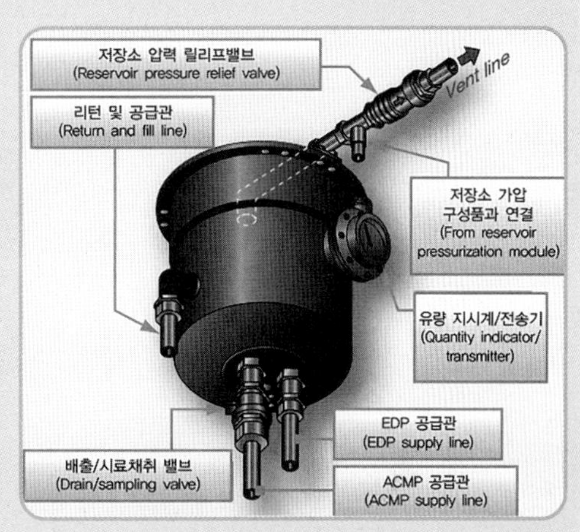

[그림 3-12] Hydraulic Reservoir

Words & Phrase

- **bleed air** 흐르는 공기
- **constant flow** 흐름의 속도가 일정한
- **crossover** 교차하는
- **foaming** 거품이 생기는 현상
- **maintain** 유지하다 (기계, 사물 등의 상태를 양호하게 유지하는 것을 의미)
- **pneumatic** 공기가 들어있는
- **prevent** 방지하다, 예방하다
- **reservoir** 저수지, 저장통

Translation

유압 리저버 가압 시스템은 비행기의 공압 시스템으로부터 여과된 에어를 메인 유압 시스템 A와 B 리저버로 공급한다. 리저버 가압 시스템으로부터의 공기압은 유압 시스템(A), 시스템(B) 및 보조 유압 시스템 리저버에 대한 압력 수두를 유지한다. 가압된 리저버들은 일정한 유체 흐름을 유압 펌프에 공급한다.

리저버는 또한 공압 시스템 크로스오버 매니폴드를 가압할 필요 없이 외부의 공기 공급 장치를 사용하여 수동으로 가압 할 수 있다. 시스템은 펌프에 유압유를 확실하게 공급하고, 유압 시스템에서 정상 리턴 압력을 유지하며, 리저버 내부에서 기포가 생기는 것을 방지하는 데 도움이 된다.

Tip

head pressure 용기에 담긴 액체의 전체 기둥이 용기 바닥에 가하는 압력으로 압력 수두라고 부른다. 저장 용기의 내부 상단에서 공압이 작용하여 액체를 아래 방향으로 누르게 된다.

1.4 Landing Gear

The most of commercial airplane has a tricycle-type landing gear with air/oil shock struts.

The landing gear structural systems consist of the main landing gear (MLG), nose landing gear (NLG) and doors. The landing gear extension and retraction systems extend and retract the main and nose landing gear.

Hydraulic system A normally supplies pressure to the landing gear extension and retraction. Hydraulic system B supplies pressure for retraction only.

The nose wheel steering system supplies the ground directional control of the airplane. Hydraulic system A for nose wheel steering comes from the nose landing gear extension pressure only.

The main landing gear also transmits the braking forces to the airplane structure.

Words & Phrase

absorb 흡수하다
extension 펼침 (비행기 L/G가 펼쳐져 있는 상태)
retraction 접힘 (비행기 L/G가 접혀있는 상태)
tricycle 바퀴가 세 개인
brake force 제동력
normally 일반적으로
transmit 전송하다

Translation

상용항공기 대부분은 공기/오일 충격 흡수장치가 있는 세발자전거 형태로 착륙 기어를 가지고 있다. 랜딩 기어의 구조는 메인 랜딩 기어(MLG), 노즈 랜딩 기어(NLG) 및 도어로 구성된다. 랜딩 기어의 펼침 및 접힘 시스템은 메인 및 노즈 랜딩 기어를 작동시킨다.

유압 시스템 A는 통상적으로 랜딩 기어의 작동을 위한 유압을 공급한다. 유압 시스템 B는 접힘만을 위한 유압을 공급한다.

노즈 휠 스티어링 시스템을 통해 지상에서 비행기의 방향을 통제할 수 있으며, 유압 시스템 A에서 랜딩 기어를 펼칠 때 같이 작동한다.

메인 랜딩기어는 비행기의 제동력을 동체에 전달하기도 한다.

1.4.1 Shock Absorbing

The two main landing gear absorbs landing forces and holds most of the airplane's weight when the airplane is on the ground.

The main landing gear shock strut uses hydraulic fluid and compressed dry air or nitrogen to control the shock strut action. The gas charging valve lets you pressurize the shock strut. The oil charging valve permits hydraulic servicing of the shock strut.

The nose landing gear absorbs landing forces and holds the forward part of the airplane's weight when the airplane makes a landing.

Words & Phrase

compressed 압축된 **landing force** 착륙할 때 가해지는 힘 **nitrogen** 질소
shock strut 충격을 흡수하는 지지대

Translation

두 개의 메인 랜딩기어는 지상에서 비행기의 무게를 지탱하고 착륙할 때 가해지는 충격을 흡수한다. 메인 랜딩 기어 쇼크 스트럿은 유압액과 압축된 건조 공기 또는 질소를 사용하여 쇼크 스트럿 작용을 제어한다. 가스 충전 밸브를 통해 질소가 공급되고, 쇼크 스트럿을 가압할 수 있게 해준다. 오일 충전 밸브를 통해서 쇼크 스트럿의 유압을 공급한다.

노즈 랜딩 기어는 착륙할 때의 충격력을 흡수하여 비행기가 착륙할 때 비행기 무게의 앞부분을 잡아준다. 노즈 기어의 스트럿은 비행기의 앞쪽을 지지한다.

[그림 3-13] Landing Gears

2) Maintenance Procedure

2.1 Hydraulic reservoir pressurization system

a. if the downlock pins are not installed in the nose and main landing gear, do Landing Gear Downlock Pins Installation

WARNING MAKE SURE THE DOWNLOCK PINS ARE INSTALLED ON ALL THE LANDING GEAR. WITHOUT THE DOWNLOCK PINS, THE LANDING GEAR COULD RETRACT AND CAUSE INJURIES TO PERSONS AND DAMAGE TO EQUIPMENT.

b. get access to the air charging valve manifold assembly in the right main landing gear wheel well.

c. make sure that the air charging valve is fully closed and a dust cap is installed on the valve stem.

d. to pressurize the pneumatic crossover manifold, supply pressure to the pneumatic system with One or Both Engines or APU or external ground air source.

e. make note of the reservoir pressure indication on the air pressure gauge in the right main landing gear wheel well, and the duct pressure indication on the bleed air pressure gauge in the flight compartment.

f. make sure that the pressure difference between the reservoir air pressure gauge and the bleed air pressure gauge is not more than 10psi.

g. make sure that the reservoir air pressure indication remains stable and does not decrease.

NOTE If the pressure decreases, there may be an air leak in the reservoir pressurization system.

Words & Phrase		
	downlock 랜딩기어 고정장치	**fully closed** 완전히 잠긴 상태
	injury 부상	**be installed in** ~에 장착되다
	make note 메모하다	**remain** 계속 ~ 이다

Translation

2.1 유압 시스템 가압

a. 랜딩기어의 다운 락 핀이 노즈 및 메인 랜딩 기어에 설치되지 않았으면 랜딩 기어 다운 락 핀을 설치한다.

WARNING 〈경고〉 모든 랜딩 기어에 다운락 핀이 설치되어 있는지 확인한다. 다운락 핀이 없으면 랜딩 기어가 접히게 되어 사람이 상해를 입거나 장비가 손상될 수 있다.

b. 우측 메인 랜딩 기어 휠에 있는 공기 충전 밸브 매니폴드 어셈블리에 접근한다.

c. 공기 충전 밸브가 완전히 닫혀 있고 밸브 스템에 더스트 캡이 설치되어 있는지 확인한다.

d. 공압 크로스오버 매니폴드에 압력을 가하려면 1개 또는 2개 엔진 또는 APU 또는 외부의 공기 공급 장치로부터 압력을 공급한다.

e. 우측 메인 랜딩 기어 휠의 공기압 게이지에 있는 리저버 압력 수치와 조종석 화면에 있는 블리딩 공기압 게이지에 있는 덕트의 압력 수치를 기록한다.

f. 탱크 공기압 게이지와 블리딩 공기압 게이지 간의 압력 차이가 10psi 이하인지 확인한다.

g. 탱크 공기압 표시가 안정적으로 유지되고 감소하지 않는지 확인한다.

참고 압력이 감소하면 탱크 가압 시스템에서 공기가 누출될 수 있다.

2.2 Landing Gear Inner Cylinder Chrome Cleaning - with Strut Extended

a. To clean as much of the chrome surface as possible, do one of these steps

1) Clean the chrome before you fill the airplane with fuel.

2) Over-inflate the shock strut with dry-air or nitrogen

3) Use airplane jacks to lift the airplane until the tires do not touch the ground

b. On the shock strut inner cylinder chrome, you may find dark smear marks that have a small amount of texture and are difficult to remove. These smear marks are created by oxidation of the self lubricating material that is deposited on the chrome by the shock strut bearings. These smear marks are considered acceptable.

CAUTION
DO NOT RUB THE SMEAR MARKS TO TRY TO REMOVE THEM.
THIS COULD CAUSE DAMAGE TO THE WEAR SURFACE.

c. Clean the dirt, oil, and other unwanted materials from the chrome surface of the inner cylinders with a clean cloth that is soaked in hydraulic fluid

d. Put the aircraft back to its usual condition.

Words & Phrase			
	as much as possible 최대한	**deposit** 침전시키다, 보증금	**dirt** 먼지
	inflate 부풀리다	**oxidation** 산화	**rub** 문지르다
	smear 번지다, 기름 자국	**soak** 담그다, 흠뻑 적시다.	**texture** 질감

Translation

2.2 랜딩 기어의 크롬 소재 이너 실린더 세척 – 스트럿 펼친 상태

a. 크롬 표면을 최대한 많이 청소하려면 다음 단계 중 하나의 절차를 수행한다.

1) 비행기에 연료를 채우기 전에 크롬 이너 실린더를 닦는다.

2) 건조한 공기 또는 질소로 실린더를 최대한 팽창 시킨다.

3) 타이어가 땅에 닿지 않을 때까지 비행기를 들어 올린다.

b. 충격 흡수장치의 이너 실린더는 크롬으로 제작되었고, 약간의 질감을 가지고 제거하기 어려운 어두운 얼룩 자국이 있다. 이러한 얼룩 자국은 스트럿 내부의 베어링에 의해 크롬에 침착되는 자체 윤활 물질의 산화에 의해 생성된다. 따라서, 이러한 얼룩 자국은 허용될 수 있다.

CAUTION 〈주의〉
모든 랜딩 기어에 다운락 핀이 설치되어 있는지 확인한다. 다운락 핀이 없으면 랜딩 기어가 접히게 되어 사람이 상해를 입거나 장비가 손상될 수 있다.

c. 이너 실린더의 크롬 표면의 먼지, 오일 및 기타 불필요한 물질을 유압액에 적신 깨끗한 천으로 씻는다.

d. 작업을 마무리하고 주변을 깨끗이 정리하여 비행기를 보통 상태로 둔다.

Tip	**usual condition** 보통 비행기 정비를 마무리한 후 주변이 정돈된 상태를 의미한다. 보통 상태로 직역할 수 있으나 작업 후 마무리하는 의미로 이해해야 한다.

3.1.3 Practice Quiz

Answer Keys p. 194

Q. Choose the correct word from the box to complete the sentences below.

reservoir	main fuel tank	nitrogen	pressurization	tricycle-type

01 The quantity of fresh air necessary for ventilation is more than for __________ .

02 It is stored in the wings of the plane and is called the ___ ______ ___ .

03 Air pressure from the __________ pressurization system maintains head pressure on hydraulic system A, system B, and the standby hydraulic system.

04 The most of commercial airplane has a __________ landing gear with air/oil shock struts.

05 The main landing gear shock strut uses hydraulic fluid and compressed dry air or __________ to control the shock strut action.

PART 3.2 Powerplant

3.2.1 Engine Structure

1) General Description

1.1 Engine Design

The engines supply thrust to the airplane. The engines also supply power to electrical, hydraulic and pneumatic of airplane systems. This is an main engine structure based on the CFM56-7 engine.

The engine has (5) sections:

(1) Fan and booster

(2) High pressure compressor (HPC)

(3) Combustor

(4) High pressure turbine (HPT)

(5) Low pressure turbine (LPT)

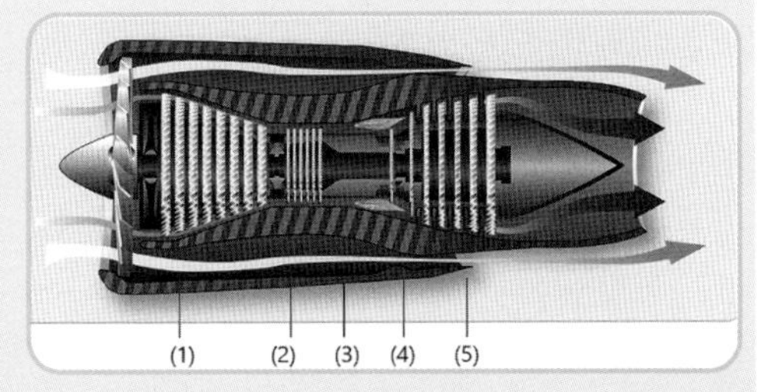

[그림 3-14] Turbo Fan Engine

Words & Phrase		
booster 승압기	**combustor** 연소실	**HPC** 고압 압축기
pneumatic 압축공기에 의한, 공압의		**thrust** 추력

Translation

엔진은 비행기에 추력을 공급한다. 엔진은 또한 비행기 시스템의 전기, 유압 및 공압에 전력을 공급하는 발전소이다. CFM56-7엔진을 기반으로 엔진의 주요 구조에 관해 설명하려 한다. 엔진의 구조를 살펴보면 아래와 같이 구성되어있다.

(1) 팬과 부스터 (2) 고압 압축기(HPC) (3) 연소실 (4) 고압 터빈 (HPT) (5) 저압 터빈 (LPT)

Tip

booster
본문에서는 승압기로 직역할 수 있으나, 가스터빈 엔진의 일반적인 구조 중 공기를 압축하는 저압 압축기 (Low Pressure Compressure, LPC)로 이해할 수 있다.

1.1.1 Fan and Compressor

The fan and booster is a four-stage compressor. The fan increases the speed of the air. A splitter fairing divides the air into primary and secondary air flows. The primary air flow goes into the core of the engine and the booster increases the pressure of this air and sends it to the HPC. The secondary air flow goes into the fan duct. It supplies approximately 80 percent of the thrust during take-off.

The HPC is a nine-stage compressor. It increases the pressure of the air from the LPC and sends it to the combustor. The HPC also supplies bleed air for the aircraft pneumatic system and the engine air system.

Words & Phrase

air flow goes to ~ 공기 흐름이 ~로 간다 **devide A into B** A를 B로 나누다
splitter 쪼개어진

Translation

팬과 부스터는 4개의 단으로 구성된 압축기이다. 팬의 역할은 공기의 속도를 증가시킨다. 스플리터 페어링은 공기를 1차 공기 흐름과 2차 공기 흐름으로 나누어준다. 1차 공기 흐름은 엔진의 중심부로 들어가고 부스터는 이 공기의 압력을 증가시켜서 HPC로 보낸다. 2차 공기 흐름은 팬 덕트로 들어간다. 팬과 부스터에서 비행기가 이륙하는 동안 대략 80%의 추력을 공급한다.

HPC는 9개의 단으로 구성된 압축기이다. LPC에서 나오는 공기의 압력을 증가시켜 연소기로 보낸다. HPC는 비행기 공압 시스템과 엔진 공기 시스템에 필요한 블리딩 공기를 공급한다.

1.1.2 Combustor and Turbine

The combustor mixes air from the compressors and fuel from the fuel nozzles. This mixture of air and fuel burns in the combustion chamber to make hot gases. The hot gases go to the HPT.

The HPT is a single-stage turbine. It changes the energy of the hot gases into mechanical energy. The HPT uses this mechanical energy to turn the HPC rotor and the accessory drive.

The LPT is a four-stage turbine. It changes the energy of the hot gases into mechanical energy. The LPT uses this mechanical energy to turn the fan and booster rotor.

Words & Phrase

burn 타오르다
chamber 공간
change into A A로 변하다 (사물/사람의 성격이 완전히 바뀔 때 사용)
mechanical 기계적인
rotor 회전자

Translation

연소기는 압축기의 공기와 연료 노즐에서 공급되는 연료를 혼합한다. 공기와 연료의 혼합물은 연소실에서 연소되어 뜨거운 가스를 만들고 이 가스는 HPT로 이동한다.

HPT는 1단 터빈이다. 연소실에서 넘어오는 뜨거운 가스 에너지를 기계적인 에너지로 바꾸는 역할을 한다. HPT는 기계적인 에너지를 사용하여 HPC 로터와 액세서리 드라이브를 회전시킨다.

LPT는 4단계로 구성된 터빈이다. HPT와 같이 뜨거운 가스의 에너지를 기계적인 에너지로 바꾸고 LPT는 팬과 부스터 로터를 작동시키기 위해 기계적인 에너지를 사용한다.

1.1.3 Engine Cowling

The engine cowling gives an aerodynamically smooth surface into and over the engine. It also gives a protective area for engine components and accessories.

There is an inlet cowl, a Fan cowl, and a thrust reverser. The inlet cowl sends air into the engine. The inlet cowl attaches to the engine. The fan cowls give an aerodynamically smooth surface over the fan case. The fan cowls attach to the fan cowl support beam. The fan cowls are open for maintenance.

Words & Phrase		
	cowl 카울 (덮개) **smooth** 부드러운	**radial** 원의 바깥쪽으로, 방사상의 **protective** 보호하는

Translation

엔진 카울은 엔진 내부와 그 위에 공기역학적으로 매끄러운 표면을 제공한다. 또한 엔진 부품과 부속품을 위한 보호 영역을 제공하는 목적도 있다. 엔진 카울의 종류로는 인렛 카울, 팬 카울 그리고 역추진 장치가 있다. 인렛 카울은 엔진으로 공기를 보내는 역할을 한다. 인렛 카울은 엔진에 장착된다. 팬 카울은 팬 케이스 위에 공기역학적으로 매끄러운 표면을 제공한다. 팬 카울은 팬 카울 지지대에 부착된다. 정비할 때 팬 카울을 열고 작업한다.

[그림 3-15] Engine Cowl

2) Maintenance Procedure

2.1 Engine Inlet and Fan Blades Inspection

a. open circuit breaker of both engine's start valve and install safety tag

b. make sure the start levers are in the CUTOFF position.

c. install a DO-NOT-OPERATE tag on the applicable start lever

d. install a protective mat in the inlet cowl

e. examine the exposed area of the spinner front cone for damage, If you find damage to the spinner front cone that is more than the limits, replace the spinner front cone.

f. do a visual check of the exposed areas of the fan blades, fan blade removal is not required.

MAKE SURE YOU WEAR GLOVES WHEN YOU HANDLE THE FAN BLADES. IF YOU DO NOT WEAR GLOVES WHEN YOU HANDLE THE FAN BLADES, YOU CAN INJURE YOUR HANDS.

g. if you find damage to the fan blade that is more than the limit, replace the fan blades.

h. put the Airplane Back to its Usual Condition

MAKE SURE THAT YOU REMOVE ALL TOOLS, PARTS OR UNWANTED MATERIAL FROM THE INLET COWL. THIS WILL PREVENT INJURY TO PERSONS AND DAMAGE TO EQUIPMENT.

i. remove protective mat from the inlet cowl, the DO-NOT-OPERATE tag from the start valve, safety tag from the circuit breaker.

j. close circuit breaker of engine's start valve.

Words & Phrase		
	applicable 해당되는	**both A and B** A와 B 둘 다
	examine 검사하다	**expose** 드러내다
	protective mat 보호 매트	**visual check** 눈으로 확인하는 점검

Translation

2.1 엔진 흡입구와 팬 블레이드 점검

a. 두 엔진 모두 스타트 밸브의 회로 차단기를 열고 안전 태그를 장착한다.

b. 엔진 스타트 레버가 컷오프 위치에 있는지 확인한다.

c. 스타트 레버에 작업 금지 태그를 설치한다.

d. 인렛 카울에 보호 매트를 설치한다.

e. 가장 앞에 있는 스피너 콘의 노출된 영역에 손상이 있는지 검사한다. 스피너 콘의 손상이 한계치 이상이면 콘을 교체한다.

f. 팬 블레이드의 노출 부위를 육안으로 확인한다. 이때, 팬 블레이드를 제거할 필요는 없다.

> **WARNING 〈경고〉** 팬 블레이드를 점검할 때 장갑을 착용한다. 날카로운 블레이드 날을 다룰 때 상해를 입을 수 있다.

g. 팬 블레이드에서 허용 한계치를 넘어서는 손상이 발견되면 팬 블레이드를 교체한다.

h. 작업을 마무리하고 주변을 정리하여 비행기를 보통 상태로 놓는다.

> **WARNING 〈경고〉** 엔진 흡입구 부근에서 모든 공구와 부품을 정리해야 한다. 장비 손상과 인명피해를 막을 수 있다.

i. 인렛 카울의 보호 매트, 스타트 밸브의 작동 금지 태그, 회로 차단기의 안전 태그를 모두 제거한다.

j. 엔진 스타트 밸브의 회로 차단기를 닫는다.

2.2 Open Fan cowl

a. when you open the inboard fan cowl panel, retract and do the deactivation procedure for the leading edge flaps.
b. release the three latches along the mating line of the fan cowl panel
c. push the trigger to release the safety catch and pull the handle to release the latch
d. manually hold the fan cowl panel and move it away from the engine until you can get access to the hold-open rods.
e. engage the hold-open rod on the engine mounted receiver.
f. hold the fan cowl panel then move the TURN/PULL sleeve in the direction of the arrow to unlock each hold-open rod. Move the fan cowl panel away from the engine until the hold-open rods extend and snap into the fully open position.

Words & Phrase

away from ~에서 떠나서 **deactivation** 비활성화 **mating line** 연결된 선
release 풀다, 느슨하게 하다 **snap** 딸깍하는 소리

Translation

2.2 엔진 팬 카울 열기

a. 팬 카울 패널을 열 때는 가장 안쪽의 인보드 플랩을 접은 다음 작동되지 않도록 적절한 조치를 취한다.
b. 팬 카울 패널의 접합 라인을 따라 아래쪽에 있는 걸쇠 3개를 해제한다.
c. 트리거를 눌러 세이프티 캐치를 해제하고 핸들을 당겨 팬 카울을 연다.
d. 수동으로 팬 카울 패널을 잡고 홀드 오픈 로드에 접근할 수 있을 때까지 엔진에서 먼 방향 쪽으로 팬 카울을 들어 올린다.
e. 엔진에 있는 마운트에 홀드 오픈 로드를 결합한다.
f. 팬 카울 패널을 잡고 화살표 방향으로 'TURN/PULL' 슬리브를 이동하여 각 홀드 오픈 로드의 잠금을 해제하면 홀드 오픈 로드가 확장된다. 완전히 열린 위치로 로드가 내려올 때까지 팬 카울을 들어 올려서 장착한다.

3.2.1 Practice Quiz

Answer Keys p. 194

Q. Choose the correct word from the box to complete the sentences below.

protective	pneumatic	gloves	secondary	mechanical

01 The engines also supply power to electrical, hydraulic and ________ of airplane systems.

02 The ________ air flow goes into the fan duct.

03 Install a ________ mat in the inlet cowl.

04 If you do not wear ________ when you handle the fan blades, you can injure your hands.

05 The HPT is a single-stage turbine. It changes the energy of the hot gases into ________ energy.

3.2.2 Engine Start

1) General Description

1.1 Engine Control

All engine fuel and control components are on the engine. The airplane fuel system supplies fuel to the engine fuel and control system. The Electronic Engine Control (EEC) controls the engine fuel and control system. Two channels in the EEC use input data to calculate the engine fuel and control outputs to operate the engine.

The airplane also gives and receives digital and analog control data to and from the engine fuel and control system. The engine fuel and control system then meters the fuel and injects it into the combustor. The engine fuel and control system also sends the necessary fuel to the engine air system so the engine operation is efficient and stable.

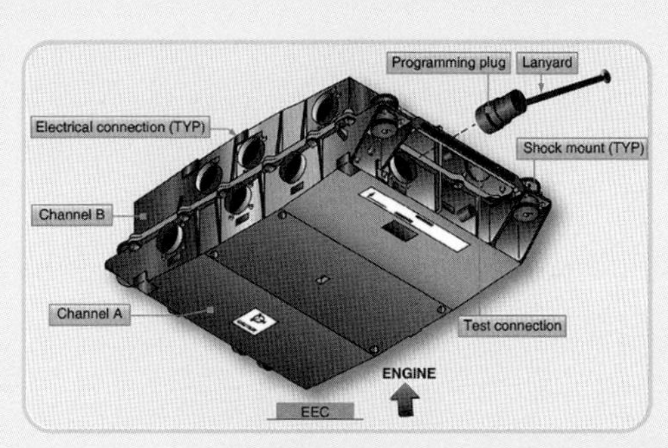

[그림 3-16] EEC

Words & Phrase			
	calculate 계산하다	**channel** 채널, 컴퓨터 동작 명령이 오가는 장치	
	commanded 지시된	**efficient** 효율적인	**inject** 주입하다
	meter 계량하다	**stable** 안정된	

Translation

모든 엔진 연료 및 제어 부품은 엔진에 있다. 비행기 연료 시스템은 엔진 연료 및 제어 시스템에 연료를 공급한다. EEC가 엔진 연료 및 제어 시스템을 제어한다. EEC의 두 채널은 입력된 데이터를 사용하여 엔진 연료를 계산하고 엔진을 작동시키기 위한 출력을 조절한다.

비행기는 엔진 연료 및 제어 시스템과 디지털 및 아날로그 제어 데이터를 주고받는다. 엔진 연료 및 제어 시스템은 연료를 측정하여 연소실에 분사하고, 필요한 연료를 엔진 공기 시스템으로 보내므로 엔진을 효율적이고 안정적으로 운영할 수 있다.

1.2 Engine Fuel

The engine fuel distribution system supplies fuel to the engine for combustion and servo system operation. The engine fuel pump assembly receives fuel from the airplane fuel system.

The ignition systems supply electrical sparks in the combustion chamber for combustion. Each engine has two ignition systems that operate independently. The ignition system usually operates manually. However, the ignition systems operate automatically when the 'EEC' sees a possible engine flameout condition.

The engine starting system operates on the ground and in flight. The engine starting system uses pneumatic power to turn turbine and compressor. Pneumatic power comes from APU, pneumatic ground equipment, or an opposite engine.

The starter turns the engine for engine starting and it changes air pressure into mechanical energy. The EEC protects the engine during the start. The EEC shuts off the fuel supply to the engine when it finds the engine parameters are out of limits during a start.

Words & Phrase

distribution 분배
ignition 점화
starter 시동장치
flameout (가스터빈엔진의) 연소 정지
parameter 변수, 요소
servo 목표에 따르도록 설정된 모터

Translation

엔진 연료 분배 시스템은 연소 및 서보 시스템 작동을 위해 엔진으로 연료를 공급한다. 비행기 연료 시스템은 엔진 연료 분배 시스템으로 연료를 공급한다. 엔진 연료 펌프는 비행기 연료 시스템으로부터 연료를 공급받는다.

점화 시스템은 연소를 위해 연소실 내에 전기 스파크를 공급한다. 각각의 엔진은 독립적으로 작동하는 2개의 점화 시스템을 갖는다. 점화 시스템은 보통 수동으로 작동한다. 그러나, EEC가 연소 정지 상태를 감지하면 자동으로 작동한다.

엔진 시동 시스템은 지상 및 비행 중에 작동한다. 압축기와 터빈을 작동시키기 위해 공압을 사용한다. 공압은 보조동력장치와 지상의 공급 장치 그리고 반대편 엔진에서 제공된다. 스타터는 공압을 기계적 에너지로 변경하여 엔진 시동을 제공한다. EEC는 시동 중에 엔진을 보호한다. 엔진 파라미터가 한계치를 벗어나면 엔진으로의 연료공급을 차단한다.

2) Maintenance Procedure

2.1 Right ignition system fail

a. open applicable circuit breakers

1)Engine No. 1 - ENG 1 IGNITION RIGHT and ENG 1 IGNITION LEFT

2) Engine No. 2 - ENG 2 IGNITION RIGHT and ENG 2 IGNITION LEFT

b. verify that start levers are in the CUTOFF position and ENGINE START switch is in the OFF position.

c. open fan cowl on the associated engine.

WARNING

MAKE SURE THE IGNITION EXCITERS ARE DE-ENERGIZED BEFORE WORKING ON THE IGNITION SYSTEM. WAIT FOR A MINIMUM OF FIVE MINUTES TO RELEASE THE HIGH VOLTAGE FROM IGNITION EXCITER.

d. disconnect LEFT/RIGHT power supply cables from the LEFT/RIGHT ignition exciter.

e. connect cables as follows:

1) LEFT power supply cable to RIGHT exciter.

2) RIGHT power supply cable to LEFT exciter.

f. close the circuit breakers opened

g. perform Audible Test of right Ignition System - EEC BITE.

h. close fan cowl on the associated engine.

Words & Phrase

applicable 해당되는
exciter 여자기
disconnect 분리하다
inoperative 이용할 수 없는
de-energized 전압이 없는 상태

Translation

2.1 오른쪽 점화 시스템 고장

a. 해당 회로 차단기를 연다.

1) 엔진 1번 – 엔진 1 점화 우측 및 엔진 1 점화 좌측

2) 엔진 2번 – 엔진 2 점화 우측 및 엔진 2 점화 좌측

b. 스타트 레버가 컷오프 위치에 있고 엔진 스타트 스위치가 OFF 위치에 있는지 확인한다.

c. 연결된 엔진의 팬 카울을 연다.

> ⚠ WARNING 〈경고〉 점화 시스템을 작동하기 전에 점화 여자기의 전원이 꺼졌는지 확인한다. 점화 여자기에서 고전압이 해제될 때까지 최소 5분 정도 기다린다.

d. 좌측/우측 점화 여자기에서 좌측/우측 전원 공급 케이블을 분리한다.

e. 다음과 같이 케이블을 교차하여 연결한다.

1) 왼쪽의 전원 공급 케이블을 오른쪽 여자기에 연결

2) 오른쪽의 전원 공급 케이블을 왼쪽의 여자기에 연결

f. 열어두었던 회로 차단기를 원위치로 닫는다.

g. 우측 점화 시스템(EEC BITE)을 통해 점화 장치의 오디오 테스트를 수행한다.

h. 연결된 엔진의 팬 카울을 닫는다.

2.2 Fuel filter removal

a. do these steps to remove the fuel filter cover:

> **NOTE** This task is for engines that have a fuel filter cover attachment with five D-Head bolts, five retaining rings, five washers, five nuts, and one bolt with its insert (1 location) into the fuel pump housing.

b. put the 5gallon (19 liters) container, under the fuel pump assembly then cut and remove the safety wire or cable from the drain plug. Let the fuel drain in the container.

c. Remove and discard the packing from the drain plug

d. loosen and remove bolt and washer of fuel filter cover

e. do an inspection of the washer for signs of damage (nick, flatness condition)

f. remove fuel filter from the fuel filter housing

g. inspect the main fuel pump housing insert for signs of damage for a pulled-out insert movement, and the thread insert damage.

h. do the inspection of the fuel filter cover and element for contamination if you find usual contamination, discard element and packing together.

Words & Phrase			
	container 수거통	contamination 오염	drain 액체를 빼낸다
	element 요소	flatness 편평함	housing 덮개
	insert 부속품	verify 확인하다	

Translation

2.2 엔진 연료 필터 버리기

a. 다음 단계를 수행하여 연료 필터 커버를 분리한다.

> **참고** 이 작업은 연료 펌프 하우징에 5개의 D-Head 볼트, 5개의 리테이닝 링, 5개의 워셔, 5개의 너트 및 1개의 볼트가 장착된 연료 필터 커버 부착물이 장착된 엔진을 대상으로 한다.

b. 5갤런(19리터) 용기를 연료 펌프 어셈블리 아래에 넣고 안전선 또는 케이블을 절단하여 드레인 플러그에서 제거한 뒤 용기에 연료를 빼낸다.

c. 패킹을 드레인 플러그에서 제거하고 버린다.

d. 연료 필터 커버의 볼트와 와셔를 풀고 분리한다.

e. 와셔에 손상 흔적이 있는지 검사한다. 손상 흔적은 찍힘과 와셔의 평탄한 정도를 파악한다.

f. 연료 필터 하우징에서 연료 필터를 분리한다.

g. 메인 연료 펌프 하우징 인서트에서 빼낸 인서트 움직임과 나사산 인서트의 손상 징후가 있는지 확인한다.

h. 일반적인 오염이 발견되면 연료 필터 커버와 엘리먼트의 오염 여부를 확인하고 패킹과 함께 버린다.

3.2.2 Practice Quiz

Answer Keys p. 194

Q. Choose the correct word from the box to complete the sentences below.

discard EEC de-energized contamination turbine

01 The ________ controls the engine fuel and control system.

02 The engine starting system uses pneumatic power to turn ________ and compressor.

03 Make sure the ignition exciters are ________ before working on the ignition system.

04 Remove and ________ the packing from the drain plug.

05 Do the inspection of the fuel filter cover and element for ________.

3.2.3 Engine Bleed Air

1) General Description

1.1 Pneumatic system

The pneumatic system supplies hot, high pressure air to the systems on the airplane that use air. There is one bleed air system for each engine. The engine bleed system controls bleed air temperature and pressure. Engine bleed air comes from the 5th and 9th stages of the high pressure compressor. A bleed air regulator and the pressure regulator and shutoff valve (PRSOV) control the flow of bleed air to the pneumatic manifold.

The EEC uses this data to control the engine air system. The EEC changes the bleed airflows to change the turbine blade tip clearances. The EEC also controls compressor airflows to prevent stalls.

Words & Phrase		
	airflow 공기 흐름	**stall** 실속
	tip clearance 블레이드 끝단의 간격	**the amount of** 어느 특정된 양

Translation

공압 시스템은 공기를 사용하는 비행기의 시스템에 고온, 고압의 공기를 공급한다. 블리드 에어 시스템은 엔진마다 하나씩 존재한다. 엔진의 블리드 시스템은 에어의 온도와 압력을 제어한다. 엔진에서 나오는 블리드 에어는 고압 압축기의 5단과 9단으로부터 나온다. 블리드 에어 레귤레이터와 압력 레귤레이터 및 차단 밸브(PRSOV)는 공압 매니폴드로의 블리드 에어의 흐름을 제어한다.

EEC는 엔진 공압 시스템을 제어하기 위해 이 데이터를 사용한다. EEC는 터빈 블레이드 팁 간극을 조절하기 위해 블리드 에어의 흐름을 변경한다. 또한, EEC는 실속을 방지하기 위해 압축공기의 흐름을 제어한다.

Tip

bleed air 엔진 압축기에서 뽑아내는 공기 흐름이며 블리드 에어는 기내 여압과 공기 순환 등에 사용된다. (교재 3.1.3 aircraft system 참조)
turbine blade tip EEC에 의해 터빈의 블레이드 간격이 제어되는데, 간격이 너무 가깝거나 멀어지면 공기 흐름에 영향을 주어 엔진 운영에 문제가 있을 수 있다.

2) Maintenance Procedure

2.1 Pneumatic duct cleaning

a. open the applicable access panels to get access to the pneumatic ducts you want to clean.

b. examine the pneumatic duct for damage. If the duct is damaged, then replace the duct.

c. clean bare titanium ducts that are not contaminated with hydraulic fluid such as manual solvent cleaner, emulsion cleaner or alkaline cleaner

d. soak a clean wiper with solvent and wring out excess solvent.

e. rub the surface with the wet wiper to remove the unwanted material.

f. wipe the duct dry with a clean wiper.

g. if bare titanium ducts with hydraulic fluid contamination, remove oil or other unwanted material.

h. clean the gold coated titanium ducts with lint-free clean cotton wiper and iso-propyl alcohol

i. put the airplane back to its usual condition and close the applicable access panels.

Words & Phrase

bare 드러내다
lint-free 보푸라기 없는
lint 부드러운 붕대용 천
wring out 액체를 꼭 짜내다

Translation

2.1 공압 덕트 청소

a. 해당 액세스 패널을 열어 청소하고자 하는 공압 덕트에 접근한다.

b. 공압 덕트의 손상 여부를 검사하고, 덕트가 손상되면 덕트를 교체한다.

c. 손으로 솔벤트 클리너, 에멀젼 클리너, 알칼리 클리너 등의 유압액에 오염되지 않은 깨끗한 티타늄 덕트를 닦는다.

d. 깨끗한 와이퍼를 필요한 클리너 용액에 담그고 여분의 용액을 제거한다.

e. 젖은 와이퍼로 표면을 문질러 오염 물질을 제거한다.

f. 깨끗한 와이퍼로 덕트를 닦고 건조한 상태를 유지한다.

g. 유압액으로 오염된 티타늄 덕트는 오일이나 기타 원치 않는 물질을 제거한다.

h. 보풀이 없는 깨끗한 면 와이퍼와 이소프로필-알코올을 이용하여 금으로 코팅된 티타늄 덕트를 닦아낸다.

i. 비행기를 보통 상태로 되돌리고 해당 액세스 패널을 닫는다.

3.2.3 Practice Quiz

Answer Keys p. 194

Q. Choose the correct word from the box to complete the sentences below.

lint-free compressor bleed

01 Engine __________ air comes from the 5th and 9th stages of the high pressure compressor.

02 The EEC also controls __________ airflows to prevent stalls.

03 Clean the gold coated titanium ducts with __________ clean cotton wiper and isopropyl alcohol

3.2.4 Engine Oil

1) General Description

1.1 Oil System

The engine oil system supplies oil to lubricate, cool, and clean the engine bearings and gears. The engine oil system has storage, distribution, and indication. The oil storage system keeps sufficient oil for a continuous supply to the oil distribution circuit.

The supply circuit sends oil to lubricate the engine bearings and gears. The lubrication unit pressurizes and filters the oil. The oil then goes to the engine. The scavenge circuit takes the oil from the engine.

The chip detectors collect and keep the unwanted materials from the scavenged oil. A chip detector has a magnet and a metallic-mesh screen. The chip detectors keep all the ferrous material or non-ferrous material pieces larger than 800 microns. This tells if there is a mechanical failure of an engine bearing or gear.

Words & Phrase

- **bearing** 베어링(물체의 회전을 지지하는 장치)
- **circuit** 회로
- **ferrous metal** 철
- **metallic-mesh screen** 금속 망
- **scavenge** 오일이 사용 후 회수되는 것을 의미
- **continuous** 계속적인
- **detect** 감지하다
- **lubricate** 윤활유를 바르다
- **micron** 100만분의 1
- **storage** 저장, 창고

Translation

엔진 오일 시스템은 엔진 베어링과 기어를 윤활, 냉각 및 세척하기 위해 오일을 공급한다. 엔진 오일 시스템은 저장, 분배 및 표시 기능이 있다. 오일 저장 시스템은 오일 분배 회로에 지속해서 공급하기 위한 충분한 오일을 탱크에 저장한다.

공급 회로는 엔진 베어링들과 기어들을 윤활하기 위해 오일을 보낸다. 탱크로부터의 오일은 누출 방지 밸브를 통해 윤활 장치로 간다. 윤활 장치는 오일을 가압하고 필터링한다. 그다음에 오일은 엔진으로 간다. 스캐빈지 회로는 엔진에 공급된 오일을 회수한다.

칩 디텍터는 배유된 오일로부터 원하지 않는 물질들을 모아서 보관한다. 칩 디텍터에는 자석과 금속-메시 스크린이 있다. 모든 철 또는 비철금속 부품이 800마이크론보다 크면 감지할 수 있다. 즉, 엔진 베어링이나 기어의 기계적인 고장이 있는지 알 수 있다.

1.2 Indicating

The oil quantity indicating system shows the engine oil quantity data on the secondary engine display. The oil quantity indicating system uses an oil quantity transmitter to measure the oil quantity in the oil tank. The oil quantity transmitter sends the oil quantity data directly to the display electronic units.

The oil quantity indicating system sends the below data to display the electronic unit.

(1) Scavenge oil filter bypass indication

(2) Low oil pressure indication

(3) Oil pressure

(4) Oil temperature

(5) Oil quantity

Words & Phrase

bypass 우회로, 우회하다
quantity (셀 수 있는) 양
directly 똑바로, 곧장
transmitter 송신기
measure 측정하다

Translation

오일량 표시 시스템은 보조 엔진 디스플레이에 엔진 오일량 데이터를 표시한다. 오일량 표시 시스템은 오일 탱크의 오일량을 측정하기 위해 오일량 송신기를 사용한다. 오일량 송신기는 오일량 데이터를 조종실의 디스플레이 장치로 직접 전송한다. 오일량 표시 시스템은 전자 장치에 표시하기 위해 아래의 데이터를 보낸다.

(1) 스캐밴지 오일 필터 바이패스 표시

(2) 저압 표시

(3) 오일 압력

(4) 오일 온도

(5) 오일량

2) Maintenance Procedure

2.1 Oil tank inspection

a. this task is the inspection procedure for the oil tank envelope
b. open fan cowl panels
c. if you find damage that is more than the limit, replace the oil tank.
d. examine the envelope for damage
  1) cracks-envelope, forward, and after bottom mount are not permitted.
  2) nicks or stretches - no more than three nicks or scratches are permitted,
    (damage does not connect)
  3) cracks or a cloudy glass window in the sight gauge are not permitted.
  4) dents or bumps - no more than three small dents or bumps are permitted.
    the damage does not distort adjacent items that you can see.
e. examine the sealing interfaces between the sight gauge and the envelope
f. put the airplane back to its usual condition and close the fan cowl panels

Words & Phrase			
adjacent 인접한, 가까운	**be permitted** 허락된다	**bump** 부딪치다, 혹, 타박상	
cloudy 흐린, 탁한	**envelope** 봉투 (위 지문에서는 오일탱크 외관으로 의역)		
gauge 게이지, 측정기	**interface** 접속하다, 접점		

Translation

2.1 오일탱크 점검
a. 이 작업은 오일 탱크 외관을 주로 검사한다.
b. 팬 카울 패널을 연다.
c. 기준치 이상의 손상이 발견되면 오일 탱크를 교체한다.
d. 오일탱크 외관의 손상 여부를 검사한다.
  1) 균열 – 봉투, 전방 및 후방 하단 마운트는 허용되지 않는다.
  2) 흠집 또는 스트레치 – 흠집이나 긁힘은 3개 이하로 허용되지 않는다. (손상이 연결되지 않음)
  3) 사이트 글래스의 균열이나 뿌옇게 흐려진 창은 허용되지 않는다.
  4) 움푹 들어간 곳이나 튀어나오는 곳 – 3개 이하의 작은 움푹 들어간 곳이나 돌기는 허용되지 않으며, 탱크 외관의 손상된 부위는 인접하지 않아야 한다.
e. 사이트 글래스 게이지와 오일 탱크 외관 사이의 밀봉 상태를 점검한다.
f. 작업을 마무리하고 주변을 정리한 뒤 팬 카울 패널을 닫는다.

3.2.4 Practice Quiz

Answer Keys p. 194

Q. Choose the correct word from the box to complete the sentences below.

lubricate isopropyl indicating scavenged

01 Clean the gold coated titanium ducts with lint-free clean cotton wiper and __________ alcohol

02 The engine oil system supplies oil to __________ cool, and clean the engine bearings and gears.

03 The chip detectors collect and keep the unwanted materials from the __________ oil.

04 The oil quantity __________ system shows the engine oil quantity data on the secondary engine display.

Electrical System

3.3.1 Electrical Power

1) General Description

1.1 Electrical Power

The electrical power system makes, supplies, and controls electrical power. The system has automatic and manual control features. Built-in test equipment (BITE) and alternate source selection make the system reliable and easy to keep.

Electrical power has these subsystems:

* Generator drive
* AC generation
* DC generation
* External power
* AC electrical load distribution.

The generator drive makes three-phase, 115/200v ac, 400Hz power for use by the electrical power system. The generator drives are the normal source of AC power in flight. The generator drive has Integrated drive generator (IDG) , Air/oil cooler and Quick attach/detach (QAD) adapter.

Words & Phrase			
	alternate 번갈아 생기는, 교대로의		**built-in** 붙박이의
	in flight 비행 중	**reliable** 신뢰할 수 있는	**supply** 공급하다

Translation

전기 전력 시스템은 전력을 만들고, 공급하고, 제어한다. 시스템은 자동으로 작동하거나 수동적으로 제어할 수 있다. 내장된 시험 장비(BITE)와 교대로 전력공급 시스템을 선택하는 것이 전기 시스템을 신뢰할 수 있고 유지하기 좋다. 비행기에 사용하는 전력에는 발전기 드라이브, 교류 전력, 직류 전력, 외부 전력, 교류 전력 분배 등의 이러한 하위 시스템이 있다.

발전기 드라이브는 3상 115/200V의 교류와 400㎐의 전력을 생산하여 전력 시스템에서 사용한다. 발전기 드라이브는 비행 중에 정상적으로 공급되는 교류 전력을 제공한다. 내부에는 통합구동 발전기와 에어 오일 쿨러 및 빠른 장착/탈착을 위한 어댑터가 있다.

[그림 3-17] Cockpit Electric Power

1.2 AC Generation

The AC generation system is a three-phase, four-wire system that operates at a nominal voltage of 115/200 volts, 400Hz. The AC system has two IDGs which are installed on engines, APU starter-generator and external power.

When pilots operate the correct flight compartment switch to use a power source. There is no automatic source selection. Priority goes to the last selection.

The AC part of the electrical system has separate left and right systems (non-parallel). This means that two power sources never supply power to the same AC transfer bus at the same time. The left and right parts of the AC systems connect if only one power supply is available.

Words & Phrase

AC generation 교류 생성
parallel 평행한
external 외부의
priority 우선 사항

Translation

교류 발전 시스템은 3상 4선 시스템으로 공칭전압인 115/200V, 400Hz로 작동한다. 교류 시스템에는 엔진에 설치되는 2개의 통합구동 발전기와 보조동력장치의 시동-발전기 그리고 외부 전력이 있다. 조종사가 동력원을 사용하기 위해서는 정확한 스위치를 작동시키면 된다. 자동으로 전력 공급원을 선택하는 기능은 없다. 마지막으로 선택한 전력 공급원이 우선순위가 된다.

전기 계통의 교류 부분은 (비평행한) 분리된 좌우 계통을 가지고 있다. 이것은 두 개의 전원이 같은 교류 이송 버스에 결코 동시에 전력을 공급하지 않는다는 것을 의미한다. 교류 계통의 왼쪽과 오른쪽 부분은 하나의 전원만 사용 가능한 경우에 연결된다.

Tip

AC generation
Alternative Current (AC) 는 교류를 뜻하며, 비행기에 주 전력원으로 사용하는 전력이다. 엔진에 장착된 IDG에서 생성하며, 교류의 사인파가 3개인 3상을 사용하여 효율이 높고 많은 전력 수요를 감당할 수 있다. 주파수를 높일수록 전선의 무게를 줄일 수 있어 항공기의 무게 경량화를 위해 400Hz를 공급한다. 해당 주파수에 이상적인 115V 전압을 공급하고 있다.

1.3 DC Generation

The DC power system is a two-wire system that operates at 28 volts (nominal). The power source for the DC system is usually the AC system. The battery supplies power if the AC system is not available. The DC system has these power sources:

* Three transformer rectifiers units (TRUs)
* Battery charger
* Battery.

The TRUs are the normal power source for the DC power system. The TRUs change 115v ac, 3-phase power to unregulated 28v dc.

The battery is a 20 cell nickel-cadmium battery with a 48 amphour capacity. With full charge, the battery gives a minimum of 30 minutes of standby AC and DC power. The battery has an internal thermal sensor. The battery charger uses this sensor to measure internal battery temperature.

Words & Phrase

available 이용할 수 있는
internal 내부의
transformer 변압기
capacity 용량
rectifier 정류기
unregulated 규제받지 않은
charger 충전기
thermal 열의

Translation

직류 전원 시스템은 28V(공칭)에서 작동하는 2선 시스템이다. 직류 시스템의 전원은 대개 교류 시스템이다. 교류 시스템을 사용할 수 없는 경우 배터리는 전원을 공급한다. 직류 공급원으로는 세 개의 변압 정류기와 배터리, 배터리 충전장치가 있다.

변압 정류기는 직류 전압의 정상 전력 공급원이다. 115V의 교류를 28V의 직류로 전환하는 역할을 한다. 배터리는 20개 셀의 니켈-카드뮴을 사용하며 48Ah 용량을 갖고 있다. 완전 충전을 하면 최소 30분의 대기 교류와 직류 전력을 공급한다. 배터리 내부에는 열 센서가 있다. 배터리충전기는 이 센서를 사용하여 내부 배터리 온도를 측정한다.

Tip

rectifier 정류기는 주기적으로 방향이 바뀌는 교류를 한 방향으로 흐르는 직류로 변환하는 장치이다.
amphour (Ah) 암페어아워는 1암페어의 전류가 1시간 동안 흘렀을 때의 전기량을 뜻하며, 충전식 배터리의 정격 용량을 나타내는 단위로 사용한다.

1.4 Standby Power System

During normal conditions, the standby power system supplies a nominal 28v dc power to battery, DC standby and switched hot battery. The standby power system supplies single phase 115v ac, 400Hz power to the AC standby bus during non-normal conditions.

The battery and standby buses give power to systems that are necessary to keep a safe flight. The standby system also gives power for ground operations when there is no AC power.

The battery supplies at least 30 minutes of AC and DC power when normal sources are not available. It supplies DC power to the DC standby bus and battery buses. The battery uses the static inverter to make AC power for the AC standby bus.

Words & Phrase

at least 최소한
bus 버스 (컴퓨터의 정보전달 회로)
hot battery 항상 배터리 전력이 공급되는 장치

Translation

정상 상태에서는 대기 전력 시스템이 배터리, 직류 스탠바이 및 스위치 핫 배터리에 공칭전압 28V의 직류 전력을 공급한다. 대기 전력 시스템은 비정상적인 조건에서 교류 전달 버스에 단상 115V 교류와 400Hz의 전력을 공급한다.

배터리와 대기 버스는 안전한 비행을 유지하는 데 필요한 시스템에 전력을 공급한다. 대기 시스템은 또한 교류 전원이 없을 때 지상 작동을 위한 전력을 공급한다.

배터리는 정상적인 소스를 사용할 수 없을 때 적어도 30분 동안의 교류 및 직류 전원을 공급한다. 직류 대기 버스와 배터리 버스에 직류 전원을 공급한다. 배터리는 정적 인버터를 사용하여 교류 대기 버스의 교류 전원을 만든다.

Tip

nominal voltage 공칭전압을 뜻하며, 전압의 종류로서 회로나 계통에 지정된 전압을 의미한다. 실제 공급되는 전압과는 차이가 있을 수 있으며, 해당 시스템을 설계할 때 지정한 전압이다.

1.5 AC Electrical Load Distribution

The AC electrical load distribution system divides the AC buses into other buses and sections. This permits better control over small electrical loads. This also protects against severe loss of power due to a single power failure.

Overcurrent load shed occurs when a generator control unit (GCU or AGCU) senses an overcurrent condition of a generator or the bus power control unit(BPCU) senses an overcurrent condition of external power. The load shed relays control power to the main buses and the galley buses to protect the power source from overload. The load shed relays open when the BPCU de-energizes them.

Words & Phrase

control over ~에 대한 통제
divide A into B A를 B로 나누다
severe 극심한, 심각한
de-energize ~의 전원을 끊다
load shed 과부하 차단

Translation

교류 전기 부하 분배 시스템은 교류 버스를 다른 버스와 섹션으로 구분한다. 이런 공급은 작은 전기 부하에 대한 더 나은 제어를 할 수 있다. 또한, 한 번의 정전으로 인한 심각한 전력 손실을 보호한다.

과전류 부하 차단은 발전기 제어 유닛이 발전기의 과전류 상태를 감지하거나 BPCU가 외부 전력의 과전류 상태를 감지할 때 발생한다. 부하 차단은 메인 버스와 갤리 버스에 제어 전력을 중계하여 과부하로부터 전원을 보호한다. 부하 차단 릴레이는 BPCU가 전원을 차단하면 열리게 된다.

Tip

load shedding 부하 차단은 전력이 과하게 공급되어 비행기의 컴퓨터나 전력 장치의 손상을 방지하기 위해 전력을 차단하는 것을 뜻한다.
galley bus 비행기의 전력을 공급하는 회로 라인 중에서 가장 처음으로 로드 쉐딩 즉, 의도적인 전력 차단을 제공하는 회로를 의미한다.

2) Maintenance Procedure

2.1 Battery Removal

a. make sure the BAT switch on cockpit overhead panel is set to the OFF position
b. get access to the forward cargo area and remove the forward bulkhead liner to remove battery.
c. disconnect the battery connector, electrical connector from the battery.
d. remove the six bolts and washers from the battery mounting brackets.
e. slide the skid plate under the battery.
f. slide the battery from the rack to the forward cargo area.

NOTE The skid plate is used so that the battery will not touch the cap strip just below it.

CAUTION

DO NOT LET THE BATTERY TOUCH THE CAP STRIP JUST BELOW THE BATTERY MOUNTING RACK. IF THE BATTERY TOUCHES THE CAP STRIP.
IT CAN SCRATCH IT AND CAUSE DAMAGE TO THE SEAL THAT FORMS WHEN THE CLOSE-OUT PANEL IS INSTALLED.

Words & Phrase		
	cap strip 뚜껑에 연결된 선	**connector** 커넥터, 전선 연결기
	mounting rack 선반 결합	**overhead** 머리 위에
	skid plate 미끄러지는 판	

Translation

2.1 배터리 장탈

a. 조종석의 윗쪽 패널에서 BAT 스위치가 꺼짐 위치로 설정되어 있는지 확인한다.
b. 전방 화물칸에 접근하여 앞부분의 벌크헤드 라이너를 제거하여 배터리를 장탈한다.
c. 배터리 커넥터, 전기 커넥터를 배터리에서 분리한다.
d. 배터리 장착 받힘대에서 6개의 볼트와 와셔를 제거한다.
e. 스키드 플레이트를 배터리 아래로 밀어 넣는다.
f. 배터리를 선반에서 앞쪽 화물칸으로 밀어 넣는다.

참고 배터리가 바로 아래쪽 캡스트립에 닿지 않도록 스키드 플레이트를 사용한다.

CAUTION
〈주의〉

배터리가 배터리 바로 아래에 있는 캡스트립에 닿지 않도록 해야 한다. 만약, 배터리 장착 선반이나 캡스트립에 배터리가 닿았을 때 패널이 긁힐 수 있으며 폐쇄 패널을 설치할 때 형성되는 밀폐실이 손상될 수 있다.

2.2 GEN 1 DRIVE light illuminated-fault isolation

a. do a check for fuel odor or fuel contamination with either the combustible gas detector unit. If you observe an overfill condition and there is fuel in the oil, replace the IDG Oil Cooler.

NOTE The IDG Oil Cooler may be leaking fuel into the Generator Oil Circuit.

b. if you do not observe fuel odor or fuel contamination, then continue.

c. if the IDG mounted to an engine is disconnected for about 50 flight hours, then replace IDG 1. If you replaced the IDG and the operational test for the IDG is satisfactory, then you corrected the problem.

d. repair confirmation do the IDG 1 operational test. If the operational test is satisfactory, then you corrected the problem.

Words & Phrase

combustible 불이 잘 붙는
contamination 오염
fault isolation 결함 분리
odor 냄새
satisfactory 만족스러운, 충분한

Translation

2.2 제너레이터 1번 구동 장치 고장 탐구

a. 가연성 가스 감지 장치를 사용하여 연료 냄새나 연료 오염을 점검한다. 과 보급 상태가 관찰되거나 오일에 연료가 있으면 IDG 오일 쿨러를 교체한다.

참고 IDG 오일 쿨러에서 발전기 오일 회로로 연료가 누출될 수 있다.

b. 연료 냄새나 연료 오염이 관찰되지 않으면 계속 다음 단계를 진행한다.

c. 엔진에 장착된 IDG가 약 50시간 비행 동안 분리되었으면 IDG 1을 교체한다. IDG를 교체하고 작동 테스트가 만족스러우면 문제가 해결된 것이다.

d. 수리 확인을 위한 IDG 1번의 작동 테스트를 수행한다. 작동 테스트가 만족스러우면 문제를 해결한 것이다.

3.3.1 Practice Quiz

Answer Keys p. 195

Q. Choose the correct word from the box to complete the sentences below.

battery generator load overfill shed

01 The _________ drive has Integrated drive generator (IDG) , Air/oil cooler and Quick attach/detach (QAD) adapter.

02 The _________ supplies power if the AC system is not available.

03 The AC electrical _________ distribution system divides the AC buses into other buses and sections.

04 The load _________ relays control power to the main buses and the galley buses to protect the power source from overload.

05 Do a check for fuel odor or fuel contamination with either the combustible gas detector unit. If you observe an _________ condition and there is fuel in the oil, replace the IDG Oil Cooler.

3.3.2 Fire Protection

1) General Description

1.1 Fire Alarms

The fire protection systems monitor the airplane for fire, smoke, overheating, and pneumatic duct leaks. The fire alarms provide visual and aural indications to the flight crew about a fire in the engines, APU, cargo compartment, and main wheel well. When there is a fire the two red FIRE WARN lights on the cockpit panel, and bell in the aural warning unit come on and the red light and horn on alternately in the right main wheel well come on (APU fire only). However, the horn does not come on in flight.

Words & Phrase			
	alternately 번갈아가며 **overheat** 과열되다	**aural** 청각의 **pneumatic** 공기압력의	**horn** 경적

Translation

화재 보호 시스템은 비행기에서 화재, 연기, 과열, 그리고 공기 중 덕트 누출을 감시한다. 화재경보기는 엔진, 보조동력장치, 화물칸, 그리고 메인 휠웰의 화재에 대해 시각적이고 청각적인 표시를 조종사에게 제공한다. 화재가 발생하면 조종석 패널에 있는 두 개의 빨간색 FIRE WARN 표시등과 청각적 경고 장치의 벨이 울리고 오른쪽 메인 휠웰의 빨간색 표시등과 경적이 번갈아 작동한다. (보조동력장치 화재만 해당). 그러나 경적은 비행 중에 울리지 않는다.

Tip

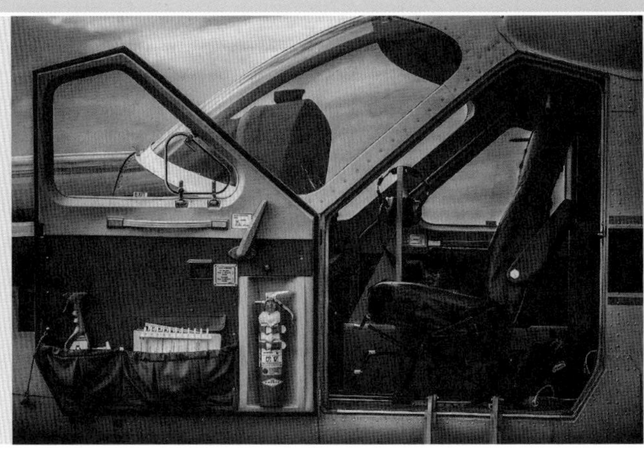

[그림 3-18] Fire Extinguisher

1.2 Engine Fire Detection

The engine fire detectors monitor for high temperatures in the engine area. Each engine has eight detectors. The detectors monitor four sections of the engine. In each section, two detectors attach to a support tube and make an assembly. An assembly has one detector from loop A and one from loop B.

The three pressure switches sense overheat, fire and fault with loss of gas pressure. Gas pressure in the sense tube holds the fault pressure switch in the closed position. The other two pressure switches close when the gas pressure increases because of an overheat or fire condition.

The overheat and fire signals go to the engine and the APU fire detection module. This module supplies overheat or fire indication in the flight compartment. If the pressure in the sense tube decreases, the fault switch opens. This switch sends the fault signal to the engine and APU fire detection module.

Words & Phrase

assembly 조립
loop 고리
detector 탐지기
loss 상실, 손실
fault 잘못, 결점
sense 감지하다

Translation

엔진 화재 감지기는 엔진 영역의 고온을 감시한다. 각 엔진에는 8개의 감지기가 있다. 감지기는 엔진의 네 부분을 감시한다. 각 부분에서 두 개의 감지기가 지지관에 부착되어 조립체를 만든다. 조립체에는 루프 A에서 한 개의 감지기와 루프 B에서 다른 한 개의 감지기가 있다.

세 개의 압력 스위치는 과열, 화재 및 가스 압력 손실로 고장을 감지한다. 감지관의 가스 압력은 고장 압력 스위치가 닫힌 위치에서 유지된다. 다른 두 개의 압력 스위치는 과열 또는 화재 상태로 인해 가스 압력이 증가하면 닫힌다.

과열 및 화재 신호는 엔진과 보조동력장치 화재 감지 모듈로 전달된다. 이 모듈은 조종석에 과열 또는 화재 표시를 공급한다. 감지 튜브의 압력이 감소하면 고장 스위치가 개방된다. 이 스위치는 고장 신호를 엔진과 보조동력장치 화재 감지 모듈로 전송한다.

Tip

fire loop 비행기의 화재 감지 시스템은 얇은 파이프 관에 가스를 채워 넣고 압축공기가 지나가는 덕트 주변이나 엔진, 보조동력장치 등 열원을 공급하는 장치 주변을 둘러싼다. 하나의 고리처럼 파이프를 연결하기 때문에 루프라고 지칭하며, 내부에 두 개의 라인이 있어 압력 차이로 인해 고온 발생을 감지한다.

1.3 Lavatory Smoke Detection

The smoke detection system gives aural and visual alarm indications when smoke is detected in a lavatory. The smoke detector is in the ceiling of each lavatory. The lavatory smoke detector connect to flight data recorder system.

The smoke sensor in each detector monitors for smoke density above a preset limit. If the density stays above the limit for appropriately eight seconds, the sensor set the detector to the alarm condition.

Words & Phrase			
	appropriately 적절하게	**ceiling** 천장	**density** 농도
	flight data recorder 비행기록장치(블랙박스)		**lavatory** 화장실
	preset 미리 결정하다		

Translation

연기 감지 시스템은 화장실에서 연기가 감지되면 청각 및 시각 경보 표시를 제공한다. 연기 감지기는 각 화장실의 천장에 장착되어 있다. 화장실의 연기 감지기는 비행 데이터 기록 시스템에 연결되어 기록된다.

각 검출기에 있는 연기 센서는 미리 설정된 한계치 이상의 연기 밀도를 감지한다. 밀도가 적절하게 8초 동안 한계치 이상으로 유지되면 센서는 검출기를 경보 조건으로 설정한다.

1.4 Overheat Detection

The wing and body overheat detection system uses sensing elements adjacent to the pneumatic ducts. It monitors the pneumatic distribution system ducts for overheat conditions. The wheel well fire detection system uses overheat sensing elements in the main wheel well. It monitors the wheel well for fire condition. When the system detects an overheat condition, alarm indications turn on in the flight compartment.

Words & Phrase			
	adjacent 인접한, 가까운	**body** 비행기 동체	**detect** 감지하다
	element 요소, 성분	**monitor** 추적 관찰하다	

Translation

날개와 동체의 과열 감지 시스템은 공압 덕트에 인접한 감지 요소를 사용한다. 공압 분배 시스템 덕트의 과열 상태를 감지한다. 휠웰의 화재 감지 시스템은 과열 감지를 사용한다.

휠웰의 화재 상태를 잘 감시한다. 시스템이 과열 상태를 감지하면, 조종실에 알람 표시가 작동한다.

1.5 Fire Extinguish

The engine fire extinguishing system floods the engine compartments with halon to put out the fire. Two fire extinguisher bottles supply the halon to either engine. The two engine fire extinguisher bottles are in the top left corner of the main wheel well. The bottles contain halon that extinguishes an engine fire.

It shapes a spherical and each bottles has halon and nitrogen at a pressure of 800psi at a temperature of 70°F (21℃).

When you pull up and turn the engine fire warning switch, you operate the squib. The squib breaks a seal in the bottle. This causes the bottle to release halon. The halon gas flows from the bottle to the selected engine compartment. If the bottle temperature increases to 266°F (130℃), the safety relief port ruptures. This releases halon into the wheel well.

Words & Phrase

extinguish 끄다
rupture 파열, 터지게 하다
squib 폭죽
flood 침수시키다, 물에 잠기다
spherical 구 모양의

Translation

엔진 소화 시스템은 불을 끄기 위해 엔진실에 할론을 가득 채운다. 두 개의 소화기 보틀이 두 엔진에 할론을 공급한다. 두 개의 엔진 소화기 보틀은 메인 휠웰의 왼쪽 위 구석에 있다.

보틀은 구의 형태이며, 내부에는 70°F(21℃)의 온도에서 800psi의 압력인 할론과 질소로 채워져 있다.

조종석에서 엔진 화재 경고 스위치를 당긴 상태로 돌리면, 소화기 보틀의 스퀴브가 작동된다. 스퀴브는 보틀 내부의 밀봉을 깨트려 할론을 방출시킨다. 할론 가스는 해당하는 엔진으로 공급된다. 보틀의 온도가 266°F(130℃)까지 올라가면, 안전 릴리프 포트가 파열되면서 휠웰 내부에 할론을 방출한다.

2) Maintenance Procedure

2.1 Lavatory smoke detection - smoke test

a. make sure that the LED status indicator on the lavatory smoke detector is a solid green light.

b. use a smoke generator to make smoke adjacent to the lavatory smoke detector.

c. make sure that the red alarm light on the lavatory smoke detector comes on.

d. make sure that the LED status indicator on the lavatory smoke detector is a solid red light.

e. make sure that you hear the smoke detector alarm horn in the lavatory.

f. make sure that these alarm indications occur in the passenger compartment:

1) the lavatory call reset switch flashes on the outside of the lavatory door

2) the attendant chime is heard through the passenger address (PA) system.

3) the amber master call light on the exit locator sign is flashing.

g. push the alarm interrupt switch on the smoke detector with an applicable tool.

h. clear the smoke from the lavatory and do the smoke test again for each of the lavatory smoke detectors.

NOTE If the smoke is not cleared from the lavatory within 60 seconds, the smoke indications will occur again.

Words & Phrase

adjacent 인접한, 가까운
interrupt 방해하다, 중단시키다, 차단하다
flash (빛으로) 신호를 보내다
status 상태, 상황

Translation

2.1 화장실 화재 감지 확인- 연기 실험

a. 화장실 연기 감지기의 상태 표시기가 녹색으로 켜져 있는지 확인한다.

b. 연기 발생기를 사용하여 화장실 연기 감지기 근처에서 연기를 만든다.

c. 화장실 연기 감지기의 빨간색 경보등이 켜져 있는지 확인한다.

d. 화장실 연기 감지기의 LED 상태 표시기가 빨간색으로 켜져 있는지 확인한다.

e. 화장실에서 연기 감지기 경보음이 들리는지 확인한다.

f. 실내에 다음과 같은 경보 표시가 나타나는지 확인한다.

1) 화장실 호출 재설정 스위치는 화장실 외부, 문 위에서 깜박거린다.

2) 승무원 차임벨은 기내 방송 시스템을 통해 들린다.

3) 출구 위치 표시판의 황색 마스터 호출 표시등이 깜박거린다.

g. 해당 도구를 사용하여 연기 감지기의 경보 차단 스위치를 누른다.

h. 화장실에서 연기를 제거하고 각 화장실 연기 감지기에 대해 연기 테스트를 다시 수행한다.

> **참고** 60초 이내에 화장실에서 연기가 제거되지 않으면 연기 표시가 다시 발생한다.

3.3.2 Practice Quiz

Answer Keys p. 195

Q. Choose the correct word from the box to complete the sentences below.

overheat	lavatory	visual and aural	halon	green

01 The fire alarms provide ___ ______ ___ indications to the flight crew about a fire in the engines, APU, cargo compartment, and main wheel well.

02 The three pressure switches sense __________ fire and fault with loss of gas pressure.

03 The smoke detection system gives aural and visual alarm indications when smoke is detected in a __________.

04 The engine fire extinguishing system floods the engine compartments with __________ to put out the fire.

05 Make sure that the LED status indicator on the lavatory smoke detector is a solid light.

3.3.3 Lights

1) General Description

1.1 Instrument and Panel Light

Many airplane systems have indicator lights in the flight compartment. Indicator lights are on instrument panels, overhead panels, electronics (aisle stand) panels. The instrument and panel lights are for the flight compartment controls and panel indications. Instrument and panel lights supply light to the switches, selectors, and indicators on panels in the flight compartment. Instruments that are necessary for flight safety connect to the standby light system. When power is not available from transfer bus 1, the lights get power from the standby bus.

The lights use incandescent lamps to supply light and the miscellaneous lights supply general lighting to the flight compartment.

The floodlights can operate in two modes, normal and standby. Not all floodlights can operate in standby mode. In the normal mode, you can adjust the intensity of the floodlight. In the standby mode, the light intensity cannot change.

Words & Phrase

aisle stand 조종석 사이에 있는 컨트롤 스탠드
incandescent 눈부시게 밝은
intensity 강도
floodlight 투광 조명(야간비행을 위한 조명등)
instrument 계기
miscellaneous 다양한, 여러 가지의

Translation

많은 비행기 시스템은 조종석에 표시등을 갖고 있다. 표시등은 계기판, 위쪽에 위치한 패널, 전자장치(조종석 스탠드) 패널에 있다. 계기판과 패널 조명은 조종석 제어 장치와 패널 표시 장치를 위한 것이다. 계기판과 패널 조명은 조종석의 스위치, 선택기 및 패널 표시 장치에 빛을 공급한다.

비행 안전을 위해 필요한 기구들이 대기 조명 시스템에 연결된다. 트랜스퍼 버스 1번에서 전원이 공급되지 않을 때는 대기 버스에서 전원을 공급받는다.

조명은 백열등을 사용하여 빛을 공급하고 다른 여러 가지 등으로 조종석에 일반 조명을 공급한다.

투광등은 일반 모드와 대기 모드의 두 가지 기능으로 작동할 수 있다. 모든 투광등이 대기 모드로 작동할 수 있는 것은 아니다. 일반 모드에서는 투광등의 세기를 조절할 수 있다. 대기 모드에서는 빛의 세기가 변할 수 없다.

1.2 Exterior Light

The exterior lighting control switches control the airplane external lights.

Landing (retractable and fixed), Runway turnoff, Logo, Position, Anti-collision, Wing, Wheel well, Taxi.

The wing illumination lights supply light to the leading edge of the wings. At night, this lets the pilots see when ice collects on the on the wing leading edges.

The landing lights help the pilots to see the runway during takeoff and landing.

The position lights show airplane position, direction and attitude to persons in other airplanes or on the ground. The position lights are red, green, and white incandescent lights. The left forward position light is red. The right forward light is green. The tail position lights are white.

The anti-collision lights make the airplane easier to see in the air and on the ground. There are five anti-collision lights on the airplane. Two red anti-collision lights and three white anti-collision lights.

The logo lights help show the airline logo or emblem on the vertical stabilizer. The lights are on the upper surface of each horizontal stabilizer.

Words & Phrase

collision 충돌
exterior 외부
landing 착륙
runway 활주로
emblem (국가 · 단체를 나타내는) 상징
external 외부의
illumination 조명
logo (회사 · 조직을 나타내는 특별히 디자인된) 상징
takeoff 이륙

Translation

외부 조명 제어 스위치는 비행기 외부의 조명을 제어한다. 착륙장치(접히거나 고정하는 형태) 고정, 활주로 턴오프, 로고, 위치, 충돌 방지, 날개, 휠웰, 활주로 이동 중에 사용한다.

날개 조명등은 날개의 앞쪽 가장자리에 빛을 공급한다. 조종사들이 야간에 얼음이 날개 앞쪽 가장자리에 모여있는지 알 수 있게 한다.

착륙등은 조종사들이 이착륙하는 동안 활주로를 볼 수 있도록 도와준다. 위치등은 비행기의 위치, 방향 그리고 자세를 다른 비행기나 지상의 사람들에게 알려준다. 위치등은 빨간색, 녹색, 그리고 흰색 백열등이 있다. 왼쪽은 빨간색으로 오른쪽은 녹색으로 동체 후방의 위치 등은 흰색을 사용한다.

충돌방지등은 비행기를 공중과 지상에서 더 쉽게 볼 수 있게 한다. 비행기에는 다섯 개의 충돌방지등이 있다. 두 개의 빨간색 충돌방지등과 흰색 충돌방지등 세 개로 구성된다.

로고등은 수직 스태빌라이저에 있는 항공사 로고나 엠블럼을 보여주는 데 도움이 된다. 조명은 각 수평 스태빌라이저의 윗면에 있다.

1.3 Interior Light

The attendant control panel (ACP) controls the passenger cabin lights. The ACP lighting display has the Passenger Seating Area, Forward Entry Area, and Aft Entry Area sections. Some buttons set the passenger cabin light intensity and color. The General Lighting Layout shows the current condition of passenger cabin light intensity and color.

The cargo and service compartment lights supply light to help maintenance personnel and ground crews. The cargo compartment lights are incandescent lamps. The lights have a micro-switch what is in the forward door frames.

Words & Phrase

attendant 객실 승무원　**interior** 내부　**maintenance** 정비
passenger 승객

Translation

객실 승무원이 사용하는 ACP는 승객 객실 조명을 제어한다. 컨트롤 패널의 조명 디스플레이에는 승객 좌석 영역, 전방 도어 영역 및 후방 도어 영역 섹션이 있다. 일부 버튼은 객실의 조명 강도와 색상을 설정한다. 일반 조명 배열은 객실 조명 강도와 색상의 현재 상태를 보여준다.

화물칸과 서비스 조명은 정비사와 지상의 직원들이 작업할 때 도움이 되는 빛을 제공한다. 화물실 조명은 백열등을 사용한다. 조명은 카고 도어 앞의 프레임에 장착된 것과 같은 마이크로 스위치 방식으로 작동한다.

[그림 3-19] Logo Light

[그림 3-20] Cabin Interior Light

1.4 Emergency Light

The emergency lighting system puts lights on areas inside and outside of the airplane. The emergency lights also show the exit paths.

The emergency lights operate when the emergency light system is on or when there is a loss of airplane DC power and the cockpit forward overhead panel emergency light switch is in the ARMED position.

The exit sign lights come on to show the location of the exits including the entry door, and aisle near the ceiling.

Words & Phrase

aisle 복도
exit path 비상탈출 경로
emergency 비상 상황
exit sign 탈출 유도등
entry 출입
include 포함하다

Translation

비상등 시스템은 비행기 내부와 외부의 영역에 조명이 있다. 비상등은 출구 경로를 보여준다. 비상등은 비상등 시스템이 켜져 있거나 비행기 직류 전원이 손실되고 조종석 전방 오버헤드 패널 비상등 스위치가 ARM 위치에 있을 때 작동한다. 출입구와 통로 근처의 천장에 비상구 안내등이 켜져서 출구 위치를 확인할 수 있다.

2) Maintenance Procedure

2.1 Wing anti-collision light assembly replacement

a. prepare for the replacement.

WARNING

DO NOT TOUCH THE ANTI-COLLISION LIGHT FOR 10 MINUTES AFTER YOU REMOVE ELECTRICAL POWER. AN ELECTRICAL SHOCK CAN CAUSE INJURIES TO PERSONNEL OR DAMAGE TO EQUIPMENT.

b. prepare a work platform or maintenance platform. These steps apply to all metal support equipment within a 50-foot radius of an open fuel tank.

c. remove the lens assembly from the anti-collision assembly.

d. remove the screws from the periphery of the lens assembly, screw, and nut to disconnect the bonding jumper from the lens assembly.

e. remove the screws from the leading edge skin cutout.

f. carefully remove the anti-collision assembly from the leading edge skin then disconnect the electrical connector.

g. prepare the electrical bonding surfaces at four electrical bonding locations common to the anti-collision assembly and leading edge skin.

h. use an intrinsically safe approved bonding meter to do a check of the electrical resistance between the leading edge skin and anti-collision assembly.

i. make sure that the electrical resistance does not exceed 0.0005 ohm

j. apply a lubricant to the faying surfaces of the lens assembly, except at electrical bonding locations. Also, let the surfaces of the lens assembly to dry.

k. apply a sealant to the external gap around the lens assembly.

Words & Phrase			
	carefully 조심스럽게	**cutout** 차단, 삭제 부분	**electrical shock** 감전
	faying surface 접합 면	**injury** 부상	**intrinsically** 본질적으로
	periphery 주변부	**radius** 반경	**work platform** 작업대

Translation

2.1 날개 충돌방지등 교체

a. 교체 작업을 준비한다.

WARNING 〈경고〉

전원을 제거한 후 10분간은 충돌방지등을 만지지 않는다. 감전으로 인해 인명피해가 발생하거나 장비가 손상될 수 있다.

b. 정비 작업을 위한 작업대를 준비한다. 이 단계는 개방형 연료 탱크에서 반경 50ft 이내에 있는 모든 금속 지지 장비에 적용된다.

c. 충돌 방지 조명 어셈블리에서 렌즈 어셈블리를 제거한다.

d. 렌즈 어셈블리 주변의 볼트와 나사, 너트를 제거하여 렌즈 어셈블리에서 본딩 점퍼를 분리한다.

e. 앞쪽의 가장자리 스킨 컷아웃에서 나사를 제거한다.

f. 앞쪽 가장자리 스킨에서 충돌 방지 어셈블리를 조심스럽게 제거한 다음 전기 커넥터를 분리한다.

g. 충돌 방지조명 어셈블리와 앞 가장자리 스킨에 공통되는 4개의 전기 결합 위치에서 전기 결합면을 준비한다.

h. 본질적으로 안전하고 승인된 접착 측정기를 사용하여 앞쪽 가장자리 스킨과 충돌 방지 어셈블리 사이의 전기 저항을 점검한다.

i. 전기 저항이 0.0005Ω을 초과하지 않는지 확인한다.

j. 전기 접합 위치를 제외하고 렌즈 어셈블리의 접착 표면에 윤활제를 바른다. 또한, 렌즈 어셈블리 표면을 건조한다.

k. 렌즈 어셈블리 주위의 외부 틈에 실런트를 바른다.

3.3.3 Practice Quiz

Answer Keys p. 195

Q. Choose the correct word from the box to complete the sentences below.

illumination paths maintenance instrument anti-collision

01 The __________ and panel lights supply light to the switches, selectors, and indicators on panels in the flight compartment.

02 The wing __________ lights supply light to the leading edge of the wings. At night, this lets the pilots see when ice collects on the on the wing leading edges.

03 The cargo and service compartment lights supply light to help __________ personnel and ground crews.

04 The emergency lighting system puts lights on areas inside and outside of the airplane. The emergency lights also show the exit __________.

05 Prepare the electrical bonding surfaces at four electrical bonding locations common to the __________ assembly and leading edge skin.

Communication & Navigation System

3.4.1 Communication

1) General Description

1.1 VHF

The very high frequency (VHF) communication system supplies communication over line-of-sight distances. It gives communication between airplanes or between ground stations and airplanes. The VHF communication system has these components:

1) Radio communication panel (RCP)
2) VHF transceiver
3) VHF antenna.

The RCP supplies selected frequency signals to tune the VHF transceivers. You can use the RCP to select the frequency of any VHF communication radio.

Words & Phrase	**communication** 통신	**distance** 거리	**line-of-sight** 시정범위(가시선)
	transceiver 트랜스시버, 송수신기		

Translation

VHF 통신 시스템은 가시거리를 통해 통신을 공급한다. 비행기 간 또는 지상국과 비행기 간의 통신을 제공한다. VHF 통신 시스템은 다음과 같은 구성 요소를 가지고 있다

1) 무선 통신 패널 2) VHF 송수신기 3) VHF 안테나

무선 통신패널은 VHF 송수신기를 조정하기 위해 선택된 주파수 신호를 공급한다. 패널을 사용하여 임의의 VHF 통신 라디오의 주파수를 선택할 수 있다.

Tip — **light-of-sight** 지상의 두 지점 간에 존재하는 자유 공간 경로를 뜻한다. 전자기파나 음파가 직선으로 전달된 것을 의미하며 무선 전송에서 송수신 안테나의 경로가 가시 선상에 있는 것을 의미하며, 경로 중간에 장애물이 없어야 한다.

1.2 ELT

The emergency locator transmitter (ELT) system automatically sends emergency signals when it senses a large change in the airplane velocity.

The flight crew can start the ELT manually at the flight deck with a switch on a control panel.

The ELT sends homing signals to search and rescue crews on the VHF and UHF emergency channels. It also sends emergency signals to satellite receivers.

The satellite receivers send this information to ground stations to calculate the location of the emergency signals. This signal also has position coordinates and airplane identification data.

Words & Phrase			
	automatically 자동으로	**calculate** 계산하다	**coordinate** 좌표
	flight deck 조종석	**homing** 유도 신호	**identification** 식별
	rescue 구하다, 구출	**satelite** 위성	**velocity** 속도

Translation

ELT 시스템은 비행기 속도의 큰 변화를 감지하면 자동으로 비상 신호를 보낸다. 조종사는 조종석에서의 수동으로 ELT를 시작할 수 있다.

ELT는 초단파와 극초단파 비상 채널의 수색 구조대에게 구조 신호를 보낸다. 위성 수신기에도 비상 신호를 보낸다. 위성 수신기들은 비상 신호의 위치를 계산하기 위해 이 정보를 지상의 기지국으로 보낸다. 이 신호에는 위치 좌표와 비행기 식별 데이터도 포함한다.

Tip

VHF와 **UHF** VHF는 지구 표면의 영향을 받으면서 전반하는 지상파 전반모드로, 주로 텔레비전방송이나 이동통신 등에 이용된다. 또 UHF는 대류권 내의 대기에 의해 영향을 받는 모드로, 높은 주파수대에서의 통신 등에 이용된다.

1.3 ACARS

The Aircraft Communications Addressing and Reporting System (ACARS) is a data-link communication system. It lets you transmit messages and reports between an airplane and an airline's ground base.

A message or report from the airplane to the airline ground base is called a down-link. A message or report from the airline ground base to the airplane is called an uplink.

ACARS automatically sends reports when necessary and at scheduled times of the flight to reduce crew workload. Typically, ACARS reports have crew identification, time information, engine performance, flight status, and maintenance items.

ACARS connects to these components of other systems:

* VHF transceiver to transmit to and receive data from the ground.

* Printer to print ACARS reports and messages.

* Remote electronics unit to distribute the chime annunciation and/or light annunciation signals.

Words & Phrase			
	annunciation 지시, 알림	**base** 기지	**be called** ~로 불리다
	distribute 나누어 주다	**transmit** 전송하다	**typically** 일반적으로, 보통
	workload 업무량, 작업량		

Translation

ACARS은 데이터 링크 통신 시스템이다. 비행기와 항공사의 지상 기지 사이에서 메시지와 보고서를 전송할 수 있다.

비행기가 항공기 지상 기지로 보내는 메시지나 보고를 다운 링크라고 한다. 항공사의 지상 기지에서 비행기로 보내는 메시지나 보고를 업링크라고 한다.

ACARS는 승무원의 업무량을 줄이기 위해, 필요할 때와 비행의 예정된 시간에 자동으로 보고서를 보낸다. 일반적으로 ACARS 보고서에는 승무원 식별, 시간 정보, 엔진 성능, 비행 상태 및 정비항목 등이 있다.

ACARS는 지상과 정보를 주고받기 위해 VHF 송수신기와 연결되며, ACARS 메시지 출력을 위한 프린터 그리고 조명과 차임으로 경보 표시 장치를 작동하기 위한 원격 전자 장치와 연결된다.

1.4 SELCAL

The selective calling (SELCAL) system supplies the flight crew with indications of calls that come in from the airline ground stations. It is not necessary for the pilots to continuously monitor company communications channels.

Airline radio networks supply communication between ground stations and airplanes. For SELCAL operation each airplane has a different four-letter code. Each letter in the code equals a different audio tone. The ground stations send the applicable tones to call an airplane.

Words & Phrase

applicable 해당하는	**come in** 밀려오다	**continuously** 연속적으로
equal to ~와 같은	**ground station** 지상 기지국	**selective** 선택적인

Translation

SELCAL 시스템은 항공사 지상국에서 들어오는 호출의 표시를 조종사에게 제공한다. 조종사가 회사 통신 채널을 지속해서 모니터링할 필요는 없다. 항공사 라디오 네트워크는 지상국과 비행기 사이에 통신을 공급한다. SELCAL의 운영을 위해 각 비행기는 다른 네 글자의 코드를 갖고 있다. 코드의 각 글자는 오디오 톤이 다른 것과 같은 의미이다. 지상국은 비행기를 부르기 위해 해당하는 톤을 보낸다.

[그림 3-21] Communication Control Panel

1.5 PA System

The passenger address (PA) system supplies this audio to the passenger cabin and flight compartment:

* Flight crew announcements
* Pre-recorded/stored announcements
* Boarding music
* Chimes.

The PA amplifier sets the priority for the audio inputs. Only one audio input signal at a time is processed. The passenger signs panel has a light that gives the ATTEND call indication. The passenger signs panel has these switches to turn on annunciations and give chimes

Words & Phrase			
	address 연설하다	**announcement** 발표	**at a time** 한 번에
	boarding music 탑승 시 나오는 음악		**cabin** 객실
	indication 지시	**input** 입력값	**processed** 가공한

Translation

PA 시스템은 이 오디오를 기내 객실과 조종실에 제공한다. PA 시스템은 조종사의 방송, 사전에 저장된 기내 방송, 탑승 시 제공되는 음악, 차임 등으로 구성된다.

PA 앰프는 오디오 입력에 대한 우선순위를 설정한다. 한 번에 하나의 오디오 입력 신호만 담당한다. 승객 사인 패널에는 승무원을 부르는 표시를 알려주는 조명이 있다. 이 패널에서 객실 승무원에게 알림을 켜고 차임을 주기 위한 스위치가 있다.

1.6 Service Interphone

The service interphone system is for flight and cabin crew and ground staff.

The flight crew selects the service interphone function from the audio control panel (ACP). Flight interphone microphones send audio to the remote electronics unit (REU). Flight interphone headsets and speakers get audio from the REU.

The flight crew can also use a handset to talk on the service interphone system.

The interphone jack connects to the system without ACP control.

The attendants operate a handset to connect to the system. An attendant panel connects the handset to the REU.

The ground crew microphones connect to the system through the service interphone switch. You must turn on the service interphone switch to operate the system from the service station jacks. The headset gets audio from the REU. The REU combines audio from the microphones, amplifies the audio signal, and sends audio to handsets, headsets, and speakers.

Words & Phrase			
	amplify 증폭시키다	**combine** 결합하다	**handset** 수화기
	headset 헤드폰	**interphone** 인터폰 (비행기 내부 통화 장치)	
	through ~을 통해	**turn on** ~을 켜다	

Translation

서비스 인터폰 시스템은 비행 및 객실 승무원과 지상 직원을 위한 것이다. 비행 승무원은 오디오 컨트롤 패널에서 서비스 인터폰 기능을 선택한다. 비행 인터폰 마이크는 오디오를 원격 전자 장치로 보낸다. 비행 인터폰 헤드셋과 스피커는 REU에서 오디오를 얻는다. 비행 승무원은 또한 서비스 인터폰 시스템에서 통화하기 위해 핸드셋을 사용할 수 있다. 인터폰 잭은 ACP 컨트롤 없이 시스템에 연결한다.

객실 승무원은 시스템에 연결하기 위해 핸드셋을 작동한다. 기내 방송패널에서 핸드셋을 REU에 연결한다

지상 직원들이 사용하는 마이크는 서비스 인터폰 스위치를 통해 시스템에 연결한다. 서비스 스테이션 잭에서 시스템을 작동하려면 서비스 인터폰 스위치를 켜야 하고, 헤드셋은 REU에서 오디오를 받는다. REU는 마이크의 오디오를 결합하고 오디오 신호를 증폭하며 핸드셋, 헤드셋, 스피커로 오디오를 송출한다.

1.7 Static discharger

There are static dischargers on the airplane to decrease radio receiver interference. The static dischargers discharge static at points as far from the fuselage as possible. This makes sure there is the least amount of coupling between the radio receiver antennas. Each discharger has a carbon fiber tip at the end of a slender rod. The rod is a resistive (conducting) material and attaches to a metal base. The base attaches and bonds to the airplane's surface. Each wing has two trailing edge dischargers. The vertical fin has a tip discharger and three trailing edge dischargers. Each side of the horizontal stabilizer has a tip discharger and two trailing edge dischargers.

Words & Phrase

at point 딱 정해진 곳
interference 간섭, 방해
static 정전기
decrease 감소하다
resistive 저항력 있는
discharger 방전자, 배출기
slender 가느다란

Translation

비행기에는 라디오 수신기 간섭을 줄이기 위해 정적 방전기가 있다. 정적 방전기는 동체로부터 가능한 한, 먼 지점에서 정적을 방전한다. 이것은 라디오 수신기 안테나들 사이의 결합량이 최소가 되도록 한다. 각각의 방전기는 가느다란 막대의 끝에 탄소 섬유의 팁을 가지고 있다. 막대는 저항성 (전도성) 물질이고 금속 베이스에 부착된다. 베이스는 기체의 표면에 붙어서 결합한다. 각각의 날개는 두 개의 트레일링 에지 방전기를 가지고 있다. 수직 핀에는 팁 방전기와 세 개의 트레일링 에지 방전기가 있다. 수평 안정기의 각 면에는 팁 방전기와 두 개의 트레일링 에지 방전기가 있다.

[그림 3-22] Interphone

[그림 3-23] Static Discharger

1.8 Voice record system

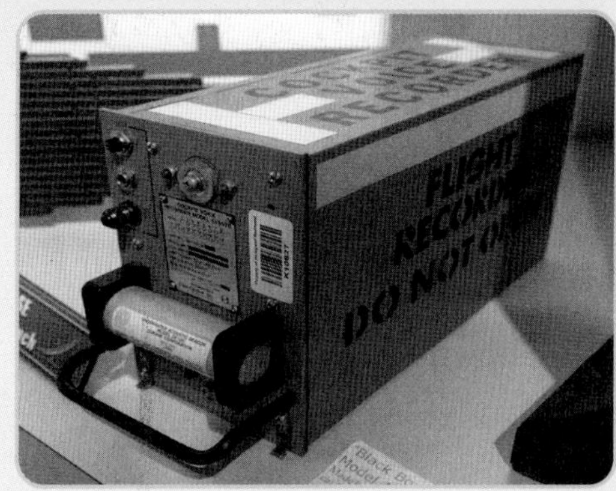

[그림 3-24] Cockpit Voice Recorder

The voice recorder continuously records these:

* Flight crew communications
* Flight compartment sounds.

The voice recorder keeps the last 120 minutes of audio.

It erases the communication data automatically so that the memory stores only recent audio. The voice recorder unit receives audio from the remote electronics unit (REU) and the area microphone. The area microphone is in the cockpit voice recorder panel.

The voice recorder unit receives time from the clock system for reference. The voice recorder unit is in the aft cargo compartment just aft of the cargo door. The container for the voice recorder unit has these properties:

* Watertight
* Shock resistant
* Heat resistant.

The voice recorder unit has an underwater locator beacon(ULB) on the front panel. The ULB helps find the voice recorder unit in water. The underwater locating device (ULD) is an ultrasonic beacon. It makes the cockpit voice recorder (CVR) easier to find if it is under water.

The ULD operates to a maximum depth of 20,000 feet (6,096 meters) and for a minimum of 30 days.

Words & Phrase			
	beacon 무선 송신소	**depth** 깊이	**recent** 최근의
	reference 언급, 참고	**resistant** 저항력 있는, ~에 잘 견디는	
	underwater 물속의	**watertight** 물이 새지 않는	

Translation

비행기의 음성 녹음기는 승무원 통신과 조종실의 지속적인 소리를 녹음한다.

음성 녹음기는 마지막 120분 동안의 오디오를 저장한다. 통신 데이터를 자동으로 소거하여 메모리가 최근의 오디오만을 저장하게 한다. 음성 녹음기 유닛은 REU와 음향 마이크로부터 오디오를 수신한다. 음향 마이크는 조종석 음성 녹음기 패널에 있다.

음성 녹음기 유닛은 참고용으로 시계 시스템으로부터 시간을 수신한다. 음성 녹음기 유닛은 화물칸 도어 바로 뒤에 있다. 음성 녹음기 유닛을 위한 컨테이너 방수, 충격에 강하고 열에 강하다.

음성 녹음기 장치는 전면 패널에 ULB를 갖고 있다. ULB는 물속에서 음성 녹음기 장치를 찾을 수 있게 한다. ULD는 초음파 비콘으로, 조종석 음성 녹음기가 물속에 있으면 더 쉽게 찾을 수 있도록 한다. ULD는 최대 수심 20,000ft(6,096미터)에서 최소 30일 동안 작동할 수 있다.

2) Maintenance Procedure

2.1 Communication systems(VHF)

a. VHF systems installed three but it is required at least two systems must be operated.

b. one may be inoperative provided VHF No. 1 operates normally.

> **NOTE** If VHF-3 is required for ATC communication, ACARS must be deactivated or VHF-3 transceiver must be swapped with the inoperative VHF-1 or VHF-2 transceiver. Dispatch is allowed with ACARS inoperative.

c. ACARS may be deactivated by opening and collaring the ACARS AC and DC circuit breakers.

d. pilot may have the difficulty to obtain weather information.

Words & Phrase

at least 적어도 **be swapped** 엇바뀌다 **deactivate** 정지시키다
inoperative 작동하지 않는

Translation

2.1 통신 시스템 - VHF

a. VHF 시스템은 3개 설치되어 있으나 최소 2개의 시스템이 작동해야 한다.

b. VHF-1이 정상적으로 작동한다면 다른 하나는 작동하지 않아도 된다.

> **참고** 항공교통관제 통신에 VHF-3이 필요한 경우 ACARS를 비활성화하거나 VHF-3 송수신기를 작동하지 않는 VHF-1 또는 VHF-2 송수신기로 교체해야 한다. ACARS가 비활성화된 상태로 비행할 수 있다.

c. ACARS는 직류와 교류 전원 공급 장치의 회로 차단기를 열고 고정하여 비활성화 시킨다.

d. 위 절차에 의해 조종사는 기상 정보를 얻는 데 어려움을 겪을 수도 있다.

2.2 Voice recorder removal

a. open these circuit breakers voice recorder

b. wait for a minimum of 11 minutes after you remove power from the voice recorder

c. to get access to the recorder independent power supply(RIPS) in the aft passenger compartment, open the lowered ceiling panel.

d. disconnect the electrical connector from the RIPS unit.

e. open the access door on the voice recorder rack

f. do this task: ESDS Handling for Metal Encased Unit Removal

CAUTION

DO NOT TOUCH THE CONNECTOR PINS OR OTHER CONDUCTORS ON THE VOICE RECORDER. IF YOU TOUCH THESE CONDUCTORS,
ELECTROSTATIC DISCHARGE CAN CAUSE DAMAGE TO THE VOICE RECORDER.

g. to remove the voice recorder do E/E Box Removal task and remove the Underwater Locator Beacon (ULB) from the voice recorder if the replacement voice recorder does not have a ULB installed.

Words & Phrase			
	compartment 공간, 칸	**electrostatic** 정전기의	**encase** 감싸다
	lowered 낮아진	**minimum** 최소한의, 최저	

Translation

2.2 음성 녹음장치 제거

a. 음성 녹음기의 회로 차단기를 연다.

b. 음성 녹음기의 전원을 끈 후 최소 11분 동안 기다린다.

c. 객실의 뒤편에서 녹음기 전력 공급 장치에 접근하려면 천장의 낮은 패널을 연다.

d. 전력 공급 장치에서 전원 커넥터를 분리한다.

e. 음성 녹음기 선반의 액세스 도어를 연다.

f. 금속으로 둘러싸인 장치 제거를 위한 정전기 제거 처리 작업을 참고하여 수행한다.

CAUTION
〈주의〉

음성녹음 장치의 커넥터 핀이나 기타 도체를 만지면 안 된다. 도체를 만지면 정전기 방전으로 인해 음성 녹음기가 손상될 수 있다.

g. 음성 녹음기를 제거하려면 E/E 박스 제거 작업을 수행하고 교체할 때 음성 녹음기에서 비콘이 설치되어 있지 않으면, ULB를 제거한다.

3.4.1 Practice Quiz

Answer Keys p. 195

Q. Choose the correct word from the box to complete the sentences below.

recorder VHF ACARS homing dischargers

01 The _________ communication system supplies communication over line-of-sight distances. It gives communication between airplanes or between ground stations and airplanes.

02 The ELT sends _________ signals to search and rescue crews on the VHF and UHF emergency channels. It also sends emergency signals to satellite receivers.

03 Typically, _________ reports have crew identification, time information, engine performance, flight status, and maintenance items.

04 There are static _________ on the airplane to decrease radio receiver interference.

05 The voice _________ continuously records flight crew communications and flight compartment sounds.

3.4.2 Navigation

1) General Description

1.1 Static and Total Air pressure

The static and total air pressure system gets air pressure inputs from three pitot probes and six static ports on the airplane fuselage.

These are the two types of air pressure:

* Static air pressure is the ambient air pressure around the airplane

* Pitot air pressure is the air pressure on the pitot probe tube as a result of the forward motion of the airplane.

Pitot probes connect to the pitot air data modules (ADMs) and the primary static ports connect to static ADMs. The ADMs change the air pressures to electrical signals and send them to the air data inertial reference units (ADIRUs). The ADIRUs use the signals to calculate flight parameters such as airspeed and altitude.

[그림 3-25] Pitot and Static Port

Words & Phrase		
	ambient air 비행기 외부의 공기	**forward motion** 전진
	fuselage 동체, 기체	**inertial** 관성의
	parameter 한도	

Translation

정압 및 전압 시스템은 비행기 동체에 있는 3개의 피토관과 6개의 정압 포트에서 공기압 입력을 얻는다. 다음 두 가지 유형의 공기압이며, 정적 공기압은 비행기 주변의 주변 기압을 뜻하고, 피토 기압은 비행기가 앞으로 가면서 인한 피토관 내부에 가해지는 기압이다. 피토 관은 피토 에어 데이터 모듈에 연결되고 기본 정적 포트는 정적 ADM에 연결된다. ADM은 공기압을 전기 신호로 변경하여 ADIRU로 보낸다. ADIRU는 신호를 사용하여 비행기의 속도와 고도 등의 비행 매개 변수를 계산한다.

Tip

pitot-static pressure 기체가 움직이는 동안, 그리고 비행을 할 때 주변에서 측정되는 공기의 압력 차이를 측정하여 비행기의 속도, 자세, 고도 등의 정보를 제공한다.
pitot pressure는 동압으로 표현하며, 항공기가 움직일 때 작용하는 압력이다. static pressure는 정압으로 해석하며, 비행기가 비행할 때 수직 방향으로 작용하는 압력을 의미한다. 두 기압의 총합을 전압(total pressure)이라고 한다.

1.2 ADIRS

The air data inertial reference system (ADIRS) has two primary functions:

* Air data reference (ADR)

* Inertial reference (IR).

The ADR function calculates airspeed and barometric altitude. The IR function calculates this data attitude, present position, groundspeed, and heading.

The TAT probe measures the outside air temperature. It changes the temperature value to an electrical signal then it goes to the ADIRUs. The AOA sensors measure and convert angle of attack to electrical signals to send to ADIRUs.

[그림 3-26] Navigation

Words & Phrase			
	airspeed 비행기의 대기 속도	**altitude** 고도	**barometer** 기압계
	convert 컨버터, 변환기	**measure** 측정하다	**primary** 주요한, 기본적인

Translation

ADIRS는 다음과 같은 두 가지 주요 기능이 있다.

* 공기 정보 참조 기준(ADR)

* 관성 기준(IR)

ADR 기능은 공기의 속도와 기압고도를 계산한다. IR 기능은 이러한 데이터 자세, 현재위치, 지상속도, 헤딩 등을 계산한다.

TAT 감지부는 외부 공기의 온도를 측정한다. 온도 값을 전기 신호로 변경한 다음 ADIRU로 보낸다.

받음각 센서는 비행기의 받음 각도를 측정하고 ADIRU로 보낼 전기 신호를 변환한다.

1.3 Standby Attitude

The standby attitude reference system is a backup system. It gives the pilots indications of airplane attitude in pitch and roll.

The standby attitude reference system also shows either the deviation from the instrument landing system (ILS) or the global landing system (GLS) localizer and glide slope deviation. It operates independently of the air data inertial reference system. The standby attitude indicator has a gyro-stabilized ball that shows horizon position to the pilots. Especially, when you need to remove power from the standby attitude indicator twenty minutes before you remove the indicator. This will give the gyro time to stop spinning.

Words & Phrase

deviation 편차
especially 특별히
independently 독립하여
either A or B A 또는 B 둘 중 하나
gyro(gyrocompass) 회전 나침반

Translation

비행기의 보조 자세계는 백업 시스템이다. 조종사들에게 비행기의 자세 표시를 피치 앤 롤로 제공한다. 보조 자세계는 ILS 또는 GLS 로컬라이저로부터의 편차 및 활공 기울기 편차를 보여준다. 이것은 항공 데이터 관성 참조 시스템과 독립적으로 작동한다. 보조 자세계에는 조종사에게 지평선 위치를 보여주는 자이로 안정화 볼이 있다. 자세계를 장탈하기 20분 전에 비행기의 전력을 제거해야 자이로가 움직임을 멈출 수 있다.

1.4 ILS

The instrument landing system (ILS) provides lateral and vertical position data necessary to put the airplane on the runway for approach. The system uses signals from a glide slope ground station and a localizer ground station.

The glide slope ground station transmits signals to give the airplane a descent path to the touchdown point on the runway. The localizer ground station transmits signals to give the airplane lateral guidance to the runway centerline.

Words & Phrase			
	approach 접근	**centerline** 활주로 중심선	**descent** 강하
	guidance 유도	**lateral** 측면의	**necessary** 필요한
	touchdown 비행기 착륙	**vertical** 수직의, 세로의	

Translation

ILS는 접근을 위해 비행기를 활주로에 올려놓는 데 필요한 측방향과 수직 방향 위치 데이터를 제공한다. 이 시스템은 지상 기지의 글라이드 슬로프 신호와 측방향의 지상 기지 신호를 이용한다.

지상의 글라이드 슬로프 신호 기지는 비행기가 활주로의 착륙 지점까지 하강 경로를 제공하도록 신호를 전송한다. 로컬라이저 지상국은 비행기가 활주로 중심선으로 좌우 측방향의 안내를 제공하기 위한 신호를 전송한다.

[그림 3-27] Localizer

1.5 Marker Beacon

The marker beacon system supplies visual and aural indications when the airplane flies over airport runway marker beacon transmitters. The marker beacon system has an antenna and a VOR/marker beacon (VOR/MB) receiver. The marker beacon antenna receives the 75MHz signal and sends it to a 75MHz filter in the VOR/marker beacon (VOR/MB) receiver 1. This filter tunes and rejects unwanted signals. If you want to hear marker beacon audio, you turn the marker beacon audio volume control.

Words & Phrase

fly over 저공비행　　**reject** 거절하다　　**tune** 음을 조정하다

Translation

마커 비콘 시스템은 비행기가 공항 활주로 마커 비콘 송신기 위를 저공 비행할 때, 시각적 및 청각적 표시를 공급한다. 마커 비콘 시스템은 안테나와 전 방향표지 시설/마커비콘 수신기로 구성된다. 마커 비콘 안테나는 75MHz의 주파수 신호를 수신하여 수신기의 필터로 전송한다. 이 필터는 원하지 않는 신호를 조정하고 제거한다. 마커 비콘 오디오를 듣고 싶다면 마커 비콘 오디오 볼륨 컨트롤러를 조절한다.

1.6 Radio Altimeter

The radio altimeter (RA) system measures the vertical distance from the airplane to the ground. The system has a range of -20 to 2,500 feet. The radio altitude shows in the flight compartment on the display units (DU). The radio altitude is computed with the receiver transmitter unit by comparing the transmitted signal to the received signal. The reflected RF signal back from the ground to determine the altitude of the aircraft. The flight crew and other airplane systems use the altitude data during low altitude flight, approach, and landing.

Words & Phrase

comparing 비교하기　　**determine** 결정하다　　**distance** 거리
range 범위　　**reflect** 반사하다

Translation

RA 시스템은 비행기로부터 지상까지의 수직 거리를 측정한다. 이 시스템은 −20ft 내지 2,500ft의 범위를 갖는다. 전파 고도는 디스플레이 유닛의 비행 구획에 표시된다. 전파 고도는 송신된 신호와 수신된 신호를 비교함으로써 수신기 송신기 유닛으로 계산된다. 지상으로부터 돌아오는 반사된 RF 신호는 항공기의 고도를 결정한다. 비행 승무원 및 다른 비행기 시스템은 저고도 비행, 접근, 착륙할 때 고도 데이터를 사용한다.

1.7 Weather Radar System

The weather radar system supplies weather conditions, wind shear events, and land contours to the flight crew on navigation displays. WXR operates on the same principle as an echo. The WXR system transmits radio frequency (RF) pulses in a 180-degree area forward of the airplane. Objects reflect the pulses to the receiver. The receiver processes the return signal to show weather, terrain, and wind shear events.

Words & Phrase			
	contour 윤곽, 등고선	**echo** 에코, 메아리, 반향	**object** 물체
	principle 원리	**terrain** 지형, 지역	**weather** 기상
	wind shear 급변풍		

Translation

기상 레이더 시스템은 날씨 상태, 급변풍 이벤트, 그리고 지상의 윤곽을 내비게이션 디스플레이를 통해서 조종사에게 제공한다. 기상 레이더는 에코와 같은 원리로 작동한다. 기상 레이더 시스템은 무선 주파수 펄스를 비행기의 전방 180도 영역에서 전송하고 물체들은 펄스를 수신기로 반사한다. 수신기는 날씨, 지형, 그리고 급변풍 이벤트를 보여주기 위해 리턴 신호를 처리한다.

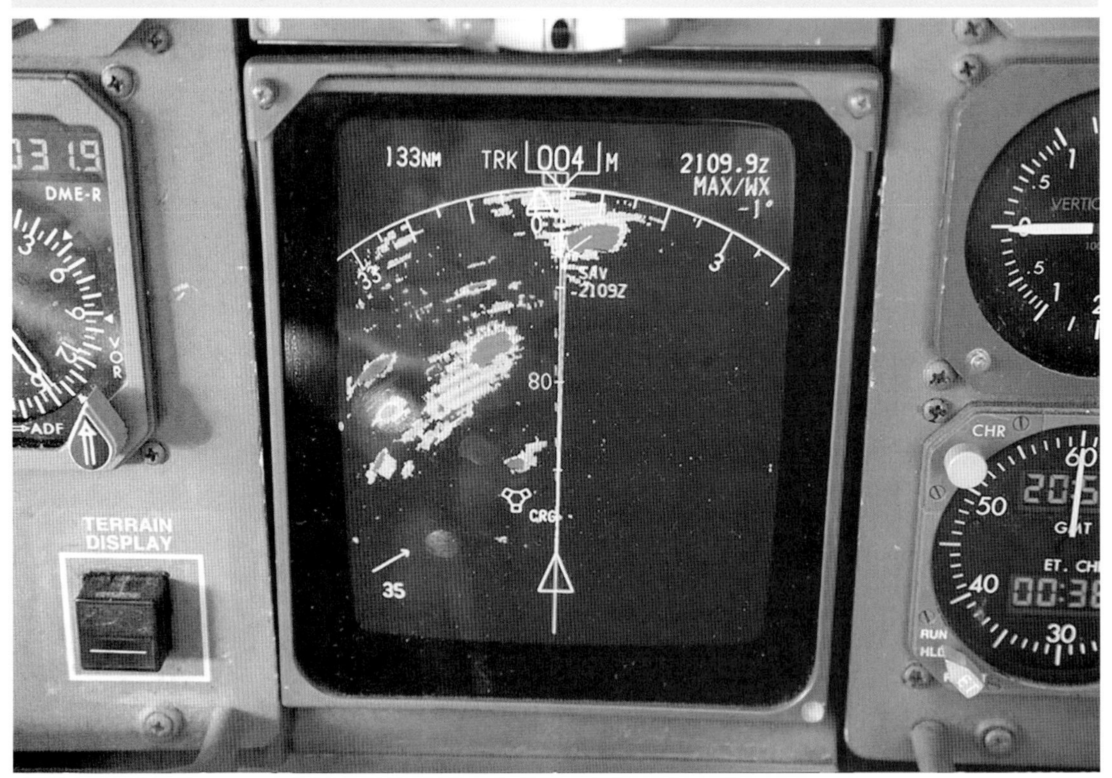

[그림 3-28] Weather Display

1.8 TCAS

The traffic alert and collision avoidance system (TCAS) helps the flight crew maintain safe air traffic separation from other ATC transponder equipped airplanes. TCAS is an airborne system and operates independently of the ground-based ATC system. TCAS sends interrogation signals to nearby airplanes.

TCAS uses these response signals to calculate the range, relative bearing, and altitude of the responding airplane. If a responding airplane does not report altitude, TCAS cannot calculate the altitude of that airplane. Airplanes tracked by TCAS are called targets.

Targets are classified as other traffic, proximate traffic, intruders, and threats. Each type of target has a different symbol on the display. TCAS also communicates with other airplanes that have TCAS to coordinate the flight movement to prevent a collision.

Words & Phrase

airborne 비행 중인, 하늘에 떠 있는 **alert** 경계하는, 알리다
avoidance 회피, 방지 **bearing** 태도, 자세, (나침반으로 측정한) 방향
classified 분류된 **collision** 충돌 **interrogate** 정보를 얻다
intruder 침입자 **nearby** 인근의 **separation** 분리

Translation

TCAS는 조종사가 다른 항공교통관제를 위한 응답장치가 장착된 비행기와 안전하게 항공 교통 분리를 유지하도록 도와준다. TCAS는 공중에서 작동하는 시스템으로 지상의 항공교통관제 시스템과 독립적으로 운영된다. TCAS는 인근 비행기에 신호를 보낸다. 이러한 응답 신호를 이용하여 대응 비행기의 사거리, 상대 비행기의 방위각, 고도를 계산한다. 만약 대응 비행기가 고도를 보고하지 않으면, TCAS는 해당 비행기의 고도를 계산할 수 없다. TCAS에 의해 추적되는 비행기를 표적이라고 한다.

표적들은 다른 트래픽, 근접 트래픽, 침입자, 및 위협들로 분류된다. 표적의 각각의 유형은 디스플레이상에서 서로 다른 심볼을 갖는다. TCAS는 또한 충돌을 방지하기 위해 비행 움직임을 조정하는 TCAS를 갖는 다른 비행기들과 통신한다.

1.9 GPWS

The ground proximity warning system (GPWS) alerts the flight crew of an unsafe condition when the airplane is near the terrain. It also supplies a warning for wind shear conditions. The GPWS uses a global positioning system (GPS) and disk-loadable software databases to give the flight crew improved terrain awareness. The system operates when the airplane is less than 2,450 feet above the ground. The GPWS also warns the flight crew of an early descent. The GPWS uses aural messages, lights, and displays to give alerts in the flight compartment.

The ground proximity warning computer (GPWC) is the main computer of GPWS and it compares the airplane flight profile, flap and gear position, and terrain clearance to find if there is a warning condition.

Words & Phrase

awareness 의식, 관심
improve 개선되다, 나아지다
proximity 가까움, 근접
disk-loadable 디스크 적재 가능한
profile 프로파일, 개요
unsafe 위험한, 불안한

Translation

GPWS는 비행기가 지형 근처에 있을 때 비행기 승무원에게 안전하지 않은 상태를 경고한다. 급변풍에 관해서도 경고 신호를 제공한다. GPWS는 조종사에게 향상된 지형 인식을 제공하기 위해 GPS와 디스크로 적재 가능한 소프트웨어 데이터베이스를 사용한다. 이 시스템은 비행기가 지상에서 2,450ft 미만 상공에 있을 때 작동한다. 또한, 조종사에게 비행기가 너무 일찍 하강할 때도 경고 신호를 보낸다.

GPWS는 조종사에 경고하기 위해 청각 메시지, 조명, 그리고 디스플레이를 사용한다. 지상 근접 경고 컴퓨터는 지상 근접 경고시스템의 메인 컴퓨터로, 비행기 비행 프로파일, 플랩 및 기어 위치, 지형의 간격 등을 비교하여 경고를 보내야 하는 상황인지 확인한다.

1.10 VOR

The VHF omnidirectional ranging (VOR) system is a navigation aid that gives magnetic bearing data from a VOR ground station to the airplane.

The VOR ground stations transmit signals that give magnetic radial information from 000 degrees to 359 degrees. All VOR stations reference the 000 degree to magnetic north.

The VOR/LOC antenna is on the top of the vertical stabilizer. The VOR antenna receives RF signals in the frequency and sends VOR signals to both VOR/MB receivers. VOR data shows on the captain and first officer displays.

Words & Phrase		
	aid 지원, 도움	**first officer** 부기장
	magnetic north 자북(자기 나침반에 나타나는 북쪽)	
	omnidirectional 모든 방향의	**radial** 방사상의

Translation

VOR시스템은 지상국에서 비행기로 자성의 방위각 데이터를 제공하는 내비게이션 보조 장치이다. VOR 지상국은 자기 방사상 정보를 000도에서 359도까지 전달하는 신호를 보낸다. 모든 VOR 지상국은 자기 북쪽으로 000도를 참조한다.

VOR/LOC 안테나는 수직 안정기의 상부에 위치한다. VOR 안테나는 주파수에서 무선 주파수 신호를 수신하고 VOR 신호를 양쪽 VOR/MB 수신기로 보낸다. 이러한 VOR 데이터는 기장과 부기장 화면에 표시된다.

1.11 GPS

The global positioning system (GPS) uses navigation satellites to supply airplane position to airplane systems and to the flight crew. The global positioning system (GPS) calculates latitude, longitude, altitude, accurate time, and ground speed.

The GPS has satellite, user, and control segments. A satellite segment is a group of satellites that orbit 10,900 nautical miles above the earth. Each satellite makes an orbit once every 12 hours. There are 21 operational satellites and 3 spares. The satellites continuously transmit radio signals with navigation data, range code, and the exact time.

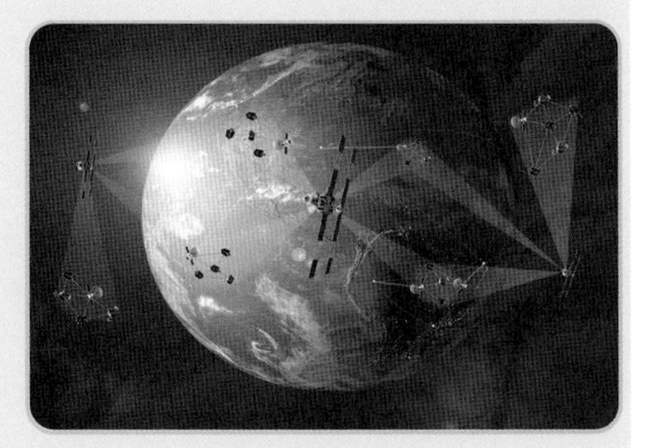

[그림 3-29] Satellites

The user segment is the GPS receiver unit on the airplane. It receives the satellite signals. The GPS uses the satellite data to calculate the airplane's position.

The control segment has control and monitor stations on Earth that continuously monitor and track the satellites. The control segment monitor and corrects satellite orbits and clocks then calculate and format a satellite navigation message. This message has up-to-date descriptions of the satellite's future positions and a collection of the latest data on all GPS satellites.

Words & Phrase

accurate 정확한
latitude 위도
satellite 위성
track 추적하다
description 서술
longitude 경도
segment 부분
up-to-date 최신의
latest 가장 최근의
orbit 궤도
spare 예비용의

Translation

GPS는 비행기 시스템과 조종사에게 비행기 위치를 제공하기 위해 항법 위성을 사용한다. GPS는 위도, 경도, 고도, 정확한 시간 및 지상 속도를 계산한다.

GPS에는 위성, 사용자 및 제어 부분으로 구성된다. 위성 부분은 지구 위 10,900해리의 궤도를 도는 위성의 그룹이다. 각 위성은 12시간마다 한 번씩 궤도를 비행한다. 21개의 작동 위성과 3개의 예비 위성이 있으며, 위성은 계속해서 항법 데이터, 범위 코드 및 정확한 시간을 가진 무선 신호를 전송한다.

사용자 부분은 비행기의 GPS 수신 장치이다. 위성 신호를 수신하며, 위성 데이터를 사용하여 비행기의 위치를 계산한다.

제어 부분은 위성을 지속해서 감시하고 추적하는 제어 및 감시 스테이션이 지구에 있다. 위성의 궤도와 시계를 감시하고 바로잡은 다음 위성 항법 메시지를 계산하고 포맷한다. 이 메시지에는 위성의 미래 위치에 대한 최신 정보와 모든 GPS 위성에 대한 최신 정보가 포함되어 있다.

2) Maintenance Procedure

2.1 ADIRU replacement

a. set the applicable mode select switch on the IRS mode select unit (MSU) to the OFF position for at least 30 seconds.

b. for the left ADIRU, open these circuit breakers.

c. to get access to the main equipment center, open electronic equipment access door.

WARNING

DO NOT REMOVE THE SHIMS FROM THE SHELF. SPECIAL EQUIPMENT AND MANUFACTURE AID IS NECESSARY TO ALIGN THE SHELF IF YOU REMOVE THE SHIMS. YOU CANNOT ALIGN THE SHELF IF YOU REMOVE THE SHIMS. IF YOU REMOVE THE SHIMS, THE AIR DATA WILL NOT BE ACCURATE, WHICH CAN CAUSE PROBLEMS WITH SAFETY OF FLIGHT.

CAUTION

DO NOT TOUCH THE CONNECTOR PINS OR OTHER CONDUCTORS ON THE ADIRU. IF YOU TOUCH THESE CONDUCTORS, ELECTROSTATIC DISCHARGE CAN CAUSE DAMAGE TO THE ADIRU.

CAUTION

REMOVAL OF THE ADIRU WILL DISABLE THE GROUND CREW CALL FUNCTION OF THE EQUIPMENT COOLING LOW FLOW DETECTION. THIS COOLING LOW FLOW CONDITION COULD CAUSE DAMAGE TO THE EQUIPMENT.

d. remove the ADIRU from the shelf

e. install the ADIRU on the shelf

f. for the left ADIRU, close these circuit breakers.

g. close electronic equipment access door.

h. do ADIRS operation test.

Words & Phrase			
	accurate 정확한, 정밀한	**align** 나란히 하다	**applicable** 해당하는
	detection 발견, 탐지	**disable** 망가뜨리다	**shelf** 선반
	shim 쐐기		

Translation

2.1 ADIRU 교체

a. 관성항법 시스템의 모드 선택 장치의 해당 모드 선택 스위치를 최소 30초 동안 OFF 위치로 설정한다.

b. 해당하는 ADIRU의 회로 차단기를 연다.

c. 메인 E/E 센터에 접근하려면 E/E 도어를 연다.

WARNING 〈경고〉 선반에서 심을 제거하지 말아야 한다. 심을 제거한 후 선반을 정렬하려면 특수 장비와 제작사의 지원이 필요하다. 심을 제거하면 선반을 정렬할 수 없다. 심을 제거하면 에어 데이터의 정확성이 없어 비행 안전에 문제가 발생할 수 있다.

CAUTION 〈주의〉 ADIRU의 커넥터 핀이나 기타 도체를 만지면 안 된다. 도체를 만지면 정전기가 발생한다. 방전으로 인해 ADIRU가 손상될 수 있다.

CAUTION 〈주의〉 ADIRU를 제거하면 장비의 냉각 저유량 감지의 지상 직원을 호출하는 기능이 비활성화된다. 냉각 흐름이 낮은 상태로 인해 장비가 손상될 수 있다.

d. 선반에서 ADIRU를 분리한다.

e. 선반에 ADIRU를 장착한다.

f. 회로 차단기를 닫는다.

g. 액세스 도어를 닫는다.

h. ADIRS 작동 점검을 수행한다.

2.2 Standby altimeter problem

a. set the BARO scale of the Standby Altimeter/Airspeed Indicator.

b. maintenance action is necessary for reports of Altitude differences between the Common Display System (CDS) and the Standby Altimeter/Airspeed Indicator.

c. above 10,000 feet and 0.4 Mach, Position Error causes the tolerance to diverge rapidly and direct crosscheck becomes inconclusive. Differences greater than 400 feet should be suspect and verified by Ground Maintenance checks.

d. if the airplane was parked in heavy rain or a high moisture environment, then do a check of the alternate Static Port.

e. if you find moisture in the Alternate Static System, do pitot static system draining task.

f. do a leak test of the Alternate Pitot and Static Systems. if the Alternate Static Systems have a leak, repair the leak.

g. flush the Alternate Pitot and then Static Systems.

h. do this check of the wiring for the Standby Altitude/Airspeed Indicator, for voltage between pin 4 of connector and ground. The correct voltage is 28 VDC.

i. do a check of continuity between pin 3 and pin 5 of connector and structure ground.

Words & Phrase

continuity 전도성, 연속성　**diverge** 갈라지다, 나누어지다　**flush** 씻어내다
inconclusive 결정적이 아닌　**moisture** 습기　**necessary** 필요한
verify 확인하다

Translation

2.2 보조 고도계 문제

a. 보조 고도계/속도계의 기압 눈금을 설정한다.

b. CDS과 대기 고도계/대기 속도 표시기 간의 고도 차이를 보고하기 위해 정비 조치가 필요하다.

c. 10,000ft 및 마하 0.4 이상에서는 위치 오류로 인해 허용 오차가 발생하여 신속하고 직접적인 교차 확인은 결론을 도출하지 못한다. 400ft보다 큰 차이는 의심되어야 하며 지상에서 정비 작업을 통해 확인해야 한다.

d. 비행기가 폭우나 습기가 많은 환경에 있는 경우에는 보조 정압 포트를 확인한다.

e. 보조 정압 시스템에서 습기를 발견하면 피토 정압 시스템의 배수 작업을 수행한다.

f. 보조 피토 및 정압 시스템의 누출 테스트를 수행한다. 보조 정압 시스템에 누출이 있는 경우 누출을 수리한다.

g. 보조 피토관을 씻은 다음 정압 시스템을 씻는다.

h. 커넥터의 핀 4와 접지 사이의 전압에 대해 보조 고도계/속도계의 전압을 확인한다. 올바른 전압은 28V 직류이다.

i. 커넥터의 핀 3과 핀 5와 구조 접지 사이의 연속성을 확인한다.

3.4.2 Practice Quiz

Answer Keys p. 195

Q. Choose the correct word from the box to complete the sentences below.

pitot bearing attitude satellite vertical

01 Pitot air pressure is the air pressure on the __________ probe tube as a result of the forward motion of the airplane.

02 It gives the pilots indications of airplane __________ in pitch and roll.

03 The radio altimeter system measures the __________ distance from the airplane to the ground. The system has a range of −20 to 2500 feet.

04 The GPS has user, satellite, and control segments. A __________ segment is a group of satellites that orbit 10,900 nautical miles above the earth.

05 The VOR system is a navigation aid that gives magnetic __________ data from a VOR ground station to the airplane.

3.4.3 Instruments

1) General Description

1.1 Display Electronic Unit

The Display Electronic Unit(DEU)s collects data from airplane systems, changes the data to a video signal to show on the display units, and sends data to other airplane systems. The DEU monitors the presence, status, and validity of inputs and cross-compares inputs with the other DEU.

The primary flight display (PFD) shows airspeed, attitude, altitude, heading, vertical speed, and flight modes with many different symbols and data.

Some indications show on the Navigation display such as ground speed, wind, true airspeed, weather radar, and TCAS data.

All of the data on the primary engine display and secondary engine display can be shown on the compacted engine display. It shows low pressure rotor speed(N1), high pressure rotor speed(N2), exhaust gas temperature(EGT), oil pressure, and fuel flow. If one of the center DUs fails, the data shows on the other center DU. The system's display shows hydraulic quantity and pressure, brake temperature, and flight control surface positions.

Words & Phrase		
	collect data 정보수집	**cross-compare** 교차 비교
	quantity 수량	**flight control surface** 비행조종면
	validity 유효함, 타당성	

Translation

DEU는 비행기 시스템으로부터 자료를 수집하고, 디스플레이 장치에 표시할 비디오 신호로 데이터를 변경하고, 다른 비행기 시스템으로 데이터를 보낸다. DEU는 입력의 존재, 상태 및 유효성을 모니터링하고 다른 DEU와 입력된 값을 교차 비교한다.

주요한 비행 정보 화면은 다양한 기호와 데이터로 비행 속도, 자세, 고도, 헤딩, 수직 속도, 비행 모드를 보여준다. 지상 속도, 바람, 실제 항공 속도, 기상 레이더 및 TCAS 데이터와 같은 일부 표시가 내비게이션 디스플레이에 표시된다.

1차 엔진 디스플레이와 2차 엔진 디스플레이의 모든 데이터를 압축하여 엔진 디스플레이에 표시할 수 있다. 저압 압축기 속도(N1), 고압 압축기 속도(N2), 배기가스 온도(EGT), 오일 압력 및 연료 유량이 표시된다. 가운데의 화면 중 하나가 고장 나면 데이터가 다른 화면에 표시된다. 시스템의 디스플레이 화면에는 유압의 양과 압력, 브레이크 온도 및 비행 조종 장치의 움직임이 나타난다.

[그림 3-30] Display Electronic Unit

1.2 Air Data Instrument

The standby altimeter/airspeed indicator is two flight instruments in one component. One instrument is a pneumatic altimeter. It gets static air pressure from the alternate static ports and shows barometric altitude.

The other instrument is a pneumatic airspeed indicator. This indicator gets pitot air pressure from the alternate pitot probe and static air pressure from the alternate static ports to show the indicated air speed.

Words & Phrase			
	alternate 번갈아 나오는	**altimeter** 고도계	**component** 부품
	standby 비상용품, 예비품		

Translation

대기 고도계/공기 속도 표시기는 하나의 부품에 있는 두 개의 비행계기다. 하나의 계기는 공압 고도계이다. 공압 고도계는 보조 정압 포트로부터 정압을 얻고 기압고도를 나타낸다. 다른 기기는 공압식 속도계이다. 속도계는 보조 피토관으로부터 피토 공기압력을 얻고 대체 정압 포트로부터 정압을 받아 표시된 에어 속도를 보여준다.

1.3 Overspeed Warning

An airplane has an airspeed and a mach limit to protect the airframe. For each aircraft type, the maximum operating speed of the aircraft is determined based on a specific altitude. The match airspeed warning system gives an aural warning that tells the flight crew when the airspeed is more than the mach or airspeed limit which is obtained from an ADIRU.

Words & Phrase			
	mach 마하수	**airframe** 기체 (엔진 제외)	**limited** 제한된
	maximum 최대	**obtain** 구하다, 얻다	

Translation

비행기는 기체를 보호하기 위해 비행 속도와 마하수의 제한이 있다. 비행기 기종에 따라 특정 고도에 따른 최대 운항 속도가 정해져있다. 비행 속도 일치 경보 시스템은 비행기의 속도가 ADIRU에서 얻은 마하 또는 비행 속도 제한 이상일 때 승무원에게 알려주는 청각적 경고를 제공한다.

1.4 Attitude Indication

The attitude area shows the pitch and the roll attitude referenced to the horizon. Pitch and roll data comes from the air data inertial reference system. In the attitude area, some indications shows flight director commands, TCAS, pitch limit, flight path vector, etc. The flight director commands show when the flight director is on. The pitch and roll commands come from the flight control computers.

Words & Phrase			
	attitude 자세계 **horizon** 수평선	**command** 명령	**flight director** 비행 지시

Translation

비행기 자세계는 지평선을 참조하여 피치와 롤 자세를 보여준다. 피치와 롤 데이터는 ADIRU에서 나온다. 자세계의 일부 표시는 비행 지시 명령, TCAS, 피치 제한, 비행경로 벡터 등을 보여준다. 비행 지시 장치는 비행 지시 표시가 켜져 있을 때를 보여준다. 피치와 롤 명령은 비행 제어 컴퓨터에서 나온다.

1.5 Altitude Indication

The altitude indication shows barometric altitude from the air data inertial reference system (ADIRS) and other related altitude information from various systems. Altitude is displayed on an altitude tape along the right side of the PFD. The altitude tape shows the barometric altitude on a moving scale. It can show a range of 806 feet. The current altitude shows in a digital readout box. The digital readout box points to the value on the altitude tape. When the captain and first officer altitudes are different by more than 200 feet, the amber ALT DISAGREE message shows at the bottom of the altitude tape.

Words & Phrase			
	barometric 기압계의 **display** (화면에) 표시 **various** 다양한	**current** 현재의 **readout** 해독, 판독	**disagree** 의견이 다르다 **scale** 범위, 눈금

Translation

고도 표시는 ADIRS의 기압고도 및 다양한 시스템으로부터의 다른 관련 고도 정보를 보여준다. 고도는 주 비행 정보 화면의 오른쪽을 따라 고도를 나타내는 띠 형태의 테이프에 표시된다. 고도계는 기압고도를 이동 눈금으로 표시한다. 고도계 표시 범위는 806ft를 나타낸다. 현재 고도는 디지털 상자에 표시된다. 디지털 상자는 테이프 형태에 나타난 값을 나타낸다. 기장과 부기장의 고도계가 200ft 이상의 차이를 지시하는 경우 황색의 고도 불일치 메시지가 하단에 표시된다.

1.6 Vertical Speed Indication

The vertical speed indication shows the vertical speed with a white pointer against the speed scale from the air data inertial reference system (ADIRS). If the vertical speed is more than 400 feet per minute, then vertical speed shows as a digital value. The digital value shows above the vertical speed scale if it is a positive vertical speed. The digital value is shown below the vertical speed scale if it is a negative vertical speed.

Words & Phrase			
	above ~보다 위에 **positive** 양의 방향	**negative** 음의 방향	**pointer** 계기판의 바늘

Translation

수직 속도 표시는 ADIRS의 속도 척도에 대해 흰색 포인터로 수직 속도를 보여준다. 수직 속도가 분당 400ft 이상이면 수직 속도는 디지털 값으로 표시된다. 디지털 값은 양의 수직 속도이면 수직 속도 척도 위에 표시된다. 디지털 값은 음의 수직 속도이면 수직 속도 척도 아래에 표시된다.

[그림 3-31] V/S and Altitude Panel

1.7 Heading Indication

The heading indication shows on a partial compass rose at the bottom of the primary flight display (PFD). The current heading shows as a triangular pointer at the top of the compass rose. The track shows as a line that extends from the center of the compass rose. The heading reference to true or magnetic north shows with a MAG or TRU indication.

Words & Phrase			
	extend 연장하다	**heading** 방향	**partial** 부분적인
	triangular 삼각형의		

Translation

방위각 표시계는 주 비행 정보 화면의 하단에 있는 부분 나침반에 표시된다. 현재의 방위는 나침반 상단의 삼각형 포인터로 나타낸다. 트랙은 나침반의 중앙에서 연장된 선으로 표시되며, 진북 또는 자북을 참고하여 자북이면 MAG으로 진북이면 TRU로 지시한다.

Tip	
	compass rose 컴퍼스 로즈로 비행장의 지면이나 시계 비행 규칙 또는 계기 비행 규칙에 그려진 방위각을 표시하는 원으로 나침반의 중앙에서 진북, 자북, 방향 등을 표시한다.

[그림 3-32] Compass Rose

2) Maintenance Procedure

2.1 Lower display unit inoperative

a. all remaining DUs operate normally.

b. it is checked that engine display can be switched to an alternate DU.

c. with the lower display unit inoperative for dispatch, depending on the failure mode, the upper display unit may exhibit flickering. Deactivating the lower display unit by opening and collaring circuit breaker.

d. verify that engine indications can be manually selected to either Captain's or First Officer's inboard DU.

Words & Phrase			
	be switched 뒤바뀌다	**collaring** 붙잡다	**exhibit** 보이다
	flickering 깜박거리는	**remaining** 남아있는	

Translation

2.1 아래 디스플레이 장치 작동하지 않을 때

a. 다른 디스플레이 유닛은 모두 정상적으로 작동해야 한다.

b. 엔진에 관련된 정보를 다른 디스플레이 유닛에서 전환할 수 있는지 확인한다.

c. 고장에 따라 아래 디스플레이 장치가 작동하지 않는 상태로 비행을 하면 위쪽의 디스플레이가 깜박거릴 수 있다. 회로 차단기를 열고 고정하여 하단 디스플레이 장치를 비활성화해야 한다.

d. 엔진 정보 표시를 기장과 부기장 자리에서 수동으로 안쪽의 디스플레이 유닛에서 선택할 수 있는지 확인한다.

2.2 Clock removal/installation

a. loosen the screws that hold the clock to the mounting clamp.

> **NOTE** The clock uses four clamp screws. There are two screws that hold the clock to the mounting clamp (large screw pair). There are two screws that hold the mounting clamp to the instrument panel (small screw pair).

b. pull the clock out of the instrument panel.

c. disconnect the electrical connector.

d. put protective covers on the electrical connector

e. remove covers and examine the connector's for bent or broken pins, dirt and damage.

f. donnect the electrical connector and put the clock into the instrument panel.

g. do the installation test to set the clock to the correct Greenwich Mean Time and date.

h. let the clock operate for not less than two minutes.

i. make sure the clock shows the correct time.

Words & Phrase

bent 구부러진 **dirt** 먼지 **examine** 조사하다, 검토하다
less than ~보다 적은 **loosen** 풀다, 느슨하게 하다

Translation

2.2 시계 장탈착

a. 시계를 장착 클램프에 고정하는 나사를 푼다.

> **참고** 시계는 4개의 클램프 나사를 사용한다. 시계를 장착 클램프(큰 나사 쌍)에 고정하는 두 개의 나사가 있고, 장착 클램프를 계기판에 고정하는 두 개의 나사(작은 나사 쌍)가 있다.

b. 계기판에서 시계를 당겨 빼낸다.

c. 전기 커넥터를 분리한다.

d. 전기 커넥터에 보호 커버를 씌운다.

e. 덮개를 제거하고 커넥터에 구부러지거나 부러진 핀, 먼지 및 손상이 있는지 검사한다.

f. 전기 커넥터를 연결하고 시계를 계기판에 넣는다.

g. 시계를 올바른 그리니치 표준시와 날짜로 설정하여 테스트를 수행한다.

h. 시계를 2분 이상 작동시킨다.

i. 시계가 정확한 시간을 표시하는지 확인한다.

Tip **GMT** 그리니치 평균시, 영국 런던의 그리니치 천문대를 기점으로 하는 협정세계시를 뜻함

3.4.3 Practice Quiz

Answer Keys p. 195

Q. Choose the correct word from the box to complete the sentences below.

video	compass	Mach	inoperative	Greenwich Mean Time

01 The Display Electronic Units collects data from airplane systems, changes the data to a ________ signal to show on the display units, and sends data to other airplane systems.

02 An airplane has an airspeed and a ________ limit to protect the airframe.

03 The heading indication shows on a partial ________ rose at the bottom of the primary flight display.

04 With the lower display unit ________ for dispatch, depending on the failure mode, the upper display unit may exhibit flickering.

05 Do the installation test to set the clock to the correct ________ ____ ____ and date.

Answer Keys

PART 3.1 Aircraft Structure

3.1.1 Structure

1. semi-monocoque
2. nacelle
3. movable
4. fail-safe
5. girt-bar

3.1.2 Wings

1. are consist of
2. attitude
3. flaps
4. increase
5. troubleshooting

3.1.3 Aircraft System

1. pressurization
2. main fuel tank
3. reservoir
4. tricycle-type
5. nitrogen

PART 3.2 Powerplant

3.2.1 Engine Structure

1. pneumatic
2. secondary
3. protective
4. gloves
5. mechanical

3.2.2 Engine Start

1. EEC
2. turbine
3. de-energized
4. discard
5. contamination

3.2.3 Engine Bleed Air

1. bleed
2. compressor
3. lint-free

3.2.4 Engine Oil

1. isopropyl
2. lubricate
3. scavenged
4. indicating

PART 3.3 Electrical System

3.3.1 Electrical Power

1. generator
2. battery
3. load
4. shed
5. overfill

3.3.2 Fire Protection

1. visual and aural
2. overheat
3. lavatory
4. halon
5. green

3.3.3 Lights

1. instrument
2. illumination
3. maintenance
4. paths
5. anti-collision

PART 3.4 Communication & Navigation System

3.4.1 Communication

1. VHF
2. homing
3. ACARS
4. dichargers
5. recorder

3.4.2 Navigation

1. pitot
2. attitude
3. vertical
4. satellite
5. bearing

3.4.3 Instruments

1. video
2. Mach
3. compass
4. inoperative
5. Greenwich mean time

PART 4
Technical English of Aircraft Maintenance Task and Records

항공기 정비 문서의 이해와 기록

PART 4에서는 항공정비사가 현장에서 수행하는 정비의 종류와 작업의 내용, 그리고 항공기의 정시 점검에 사용하는 작업 카드의 구성과 실제 영어 문장들을 살펴보고 간편 기술 영어(STE)의 작성 기준을 적용한 정비 기록의 작성 요령에 대해 학습하고자 한다.

4.1 Aircraft Maintenance
4.2 Maintenance Program
4.3 Maintenance Task Card
4.4 Maintenance Records

PART 4.1 Aircraft Maintenance

4.1.1 Aircraft Maintenance Category

Aircraft maintenance is divided into the maintenance that is performed routinely and that is not performed routinely for the operation of the aircraft. And the maintenance performed on a routine basis is further classified as follows depending on whether it is performed repeatedly or not.

Words & Phrase		
	is (be) classified as ~ ~와 같이 분류되다	**depending on ~** ~에 따라
	is (be) divided into ~ ~로 나누어지다 (~로 구분되다)	
	further 다시, 더 나아가	**on a routine basis** 일상적으로
	perform 실시(수행)하다	**repeatedly** 반복적으로
	routinely 일상적으로	**whether ~ or not** ~인지 아닌지 (~여부)

Translation

항공기의 정비는 항공기의 운항을 위해 일상적으로 수행하는 정비와 일상적으로 수행하지 않는 정비로 구분한다. 일상적으로 수행하는 정비는 반복 수행 여부에 따라 다시 다음과 같이 분류한다.

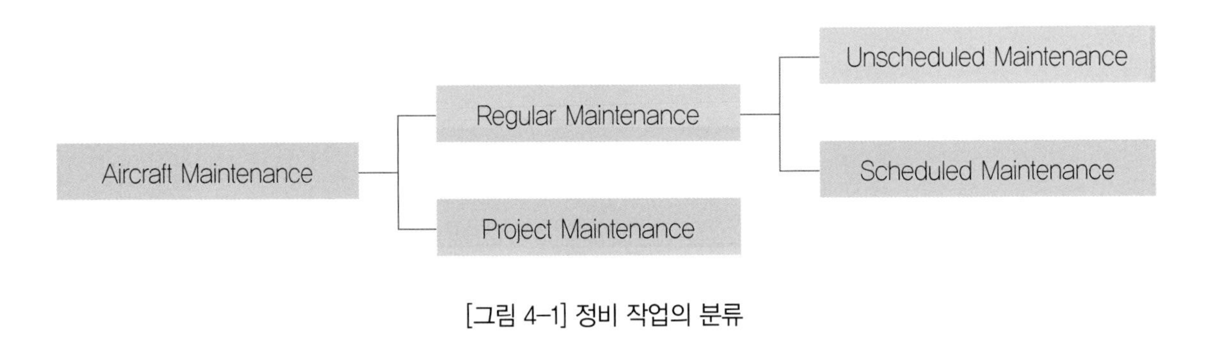

[그림 4-1] 정비 작업의 분류

1) Regular Maintenance

The Regular Maintenance is performed routinely for aircraft operation and includes Unscheduled Maintenance and Scheduled Maintenance.

Words & Phrase

aircraft operation 항공기 운항(운용)　**include(s) ~** ~을 포함하다
~ is(be) performed routinely ~이(은) 일상적으로 실시(수행)되다
scheduled maintenance 계획 정비　**unscheduled maintenance** 비계획 정비

Translation

정규 정비는 항공기의 운항을 위해 일상적으로 수행되는 것으로 비계획 정비와 계획 정비를 포함한다.

2) Project Maintenance

The Project Maintenance is only performed in a special cases and includes the maintenance such as Service Bulletin, Airworthiness Directives, Performance Improvement Modification, Pax-to-Freighter Conversion work and others.

Words & Phrase

Airworthiness Directives(AD) 감항성 개선지시　**in a special case** 특별한 경우에
Pax-to-Freighter Conversion 화물기 변환 작업
Performance Improvement Modification 성능향상 개조 작업
Service Bulletin(SB) 정비개선 회보

Translation

기획 정비는 특별한 경우에만 수행하는 것으로 정비개선 회보, 감항성 개선지시, 성능향상 개조 작업, 화물기 변환 같은 정비 작업 등을 포함한다.

3) Unscheduled Maintenance

The Unscheduled Maintenance is performed unexpectedly to repair malfunctions that occur during aircraft operation or to check for abnormalities in the aircraft when the aircraft experienced abnormal conditions. In most cases, this is not performed repeatedly. This Maintenance includes Special Inspections such as Hard Landing Inspection and Lightning Strike Inspection.

Words & Phrase

abnormal conditions 비정상적인 상태
experience ~을 경험하다, ~을 겪다
in most cases 대부분은
malfunction 기능 불량
repair 수리하다, 고치다
special inspection 특별 검사(점검)
abnormality 이상(비정상)
hard landing 경착륙
lightning strike 낙뢰(벼락을 맞음)
occur 발생하다
repeatedly 반복하여, 되풀이하여
unexpectedly 예기치 않게, 뜻밖에

Translation

비계획 정비는 항공기 운항 중에 발생한 고장을 수리하거나 항공기가 비정상적인 상태를 겪었을 때 항공기에 이상이 발생했는지 확인하기 위해 불시에 수행하는 것으로 대부분은 반복적으로 수행하지 않는다. 비계획 정비는 경착륙 검사, 낙뢰 검사 등과 같은 특별 검사를 포함한다.

4) Scheduled Maintenance

The Scheduled Maintenance is performed repeatedly at regular intervals according to the maintenance program, regardless of whether the aircraft has defects or not. This Maintenance includes various flight checks and periodic checks that are conducted at certain times and intervals for aircraft operation.

Words & Phrase

according to ~ ~에 따라
at regular intervals 일정한 간격(주기)으로
defect 결함
interval (시간적인) 간격, 주기, 시간대
periodic check 주기(정시) 점검
repeatedly 반복하여, 되풀이하여
at certain time 특정한 시간(시기)에
conduct (수행)하다
flight check 운항 점검(비행을 위한 사전 점검)
maintenance program 정비 프로그램
regardless of ~ ~에 관계없이
various 다양한

Translation

계획 정비는 항공기의 결함 유·무에 관계없이 정비 프로그램에 따라 일정한 주기로 반복적으로 수행한다. 계획 정비는 항공기 운항을 위해 일정한 시기와 주기로 실시되는 각종 운항 점검과 정기 점검을 포함한다.

4.1.2 Scheduled Maintenance Category

Scheduled Maintenance is divided into the Flight Check and the Periodic Check.

Words & Phrase		
	is (be) divided into ~ ~로 나누어지다 (~로 구분하다)	
	flight check 운항(비행) 점검	**periodic check** 주기(정시) 점검

Translation

계획 정비는 운항 점검과 주기 점검으로 나눈다.

1) Flight Check

This check is performed at regular intervals for the operation of an aircraft and is classified into several types depending on the time, scope, and intervals. This check, in most case, are performed by using checklists at an operational site of aircraft. Aircraft manufacturers may call these checks by different names.

Words & Phrase		
	aircraft manufacturer 항공기 제작사	**by different name** 다른 이름으로
	by using checklist 점검표를 사용하여	**call** ~로 부르다
	in most case 대부분은	**is (be) classified into ~** ~로 분류된다
	operational site 운항 현장	**scope** 범위

Translation

이 점검은 항공기의 운항을 위해 일정한 주기로 수행하는 점검으로 수행 시기, 범위, 주기에 따라 여러 종류로 나눈다. 이 점검은, 대부분은 점검표를 사용하여 항공기 운항 현장에서 수행한다. 항공기 제작사는 이러한 점검을 서로 다른 이름으로 부르기도 한다.

❶ Pre/Post Flight Check

This is a check to confirm the overall readiness of aircraft for resumption of flight. It is performed once between the last flight of the day and the first flight of the next day to check the overall condition of the aircraft, confirm corrective actions for any defects that have occurred, and replenish various consumable materials.

Words & Phrase	
confirm ~을 확인하다	**consumable materials** 소모성 물질 (연료, 작동유, 오일, 물,…)
corrective action (결함) 수정 작업 (발생한 결함을 해소하는 작업)	
first flight 최초 비행	**last flight** 최종 비행
occur 발생하다	**overall** 전반적인
readiness of ~ ~을 위한 준비(상태)	**replenish** 보충하다
resumption 재개(시)	

Translation

이 점검은 항공기의 비행 재개에 필요한 항공기의 전반적인 준비 상태를 확인하는 점검이다. 그날의 최종 비행과 다음 날의 최초 비행 전 사이에 한 차례 수행하며 항공기의 전반적인 상태를 점검하고, 발생한 결함에 대한 조치를 확인하며 각종 소모성 물질을 보충한다.

❷ Transit Check

This is a check performed at the departure base and intermediate base whenever the flight changes. Check the condition of the aircraft for departure and replenish fuel and consumable liquids such as engine oil.

Words & Phrase	
consumable liquids 소모성 액체	**departure base** 출발 기지
intermediate base 중간 기지	**replenish** 보충하다
whenever~ ~할 때마다	

Translation

이 점검은 항공기 출발 기지 및 중간 기지에서 운항 편이 바뀔 때마다 실시하는 점검이다. 항공기의 출발을 위한 항공기의 상태를 점검하고 연료 및 엔진 오일과 같은 소모성 액체를 보충한다.

❸ Weekly Check (주간 점검)

This is a check performed every 7 days to check the overall condition of the aircraft as well as internal and external damage, leaks, wear, and others. However, this type of check is not adopted by all of the aircraft manufacturers.

Words & Phrase		
aircraft manufacturer 항공기 제작사		**adopt** 채택하다
external 외부의	**however** 하지만, 그러나	**internal** 내부의
overall 전반적인	**as well as ~** ~뿐만 아니라	

Translation

이 점검은 항공기의 내부 및 외부의 손상, 누설, 마모 등과 전반적인 상태를 확인하기 위해 7일마다 수행하는 점검이다. 그러나 이 점검 방식을 모든 항공기 제작사가 채택하고 있지는 않다.

2) Periodic Check

The Periodic checks are preventive maintenance and airworthiness checks for various systems and structural parts of the aircraft.
These checks are performed by using various kind of task cards and are repeated at the intervals determined by flight hours, flight cycles, or days/months or years.
Manufacturers provide the periodic check requirements as individual items, and Air Carriers can group the items in the form of "A," "C," or "D" checks or in specific groups at their convenience.
Normally, "A" check is regarded as line(operational) maintenance and can be performed at operational site(ramp area), but "C" and "D" inspection are heavy maintenance and are performed inside maintenance facilities such as hangars.

Words & Phrase

air carrier 항공운송회사 (여객/화물을 나르는 항공사의 총칭)
airworthiness check 감항성 점검(확인)
at one's convenience ~의 편의대로
determined by~ ~로 정해진
facility 시설
flight cycles 비행 횟수
flight hours 비행시간
group 묶다, 모으다, 집단으로 나누다
hangar 격납고(항공기 창고)
heavy maintenance 중 정비(↔경 정비, 운항 정비)
individual 개별의 (각각의)
in the form of ~ ~의 형태로
line maintenance 운항 정비
normally 보통, 통상적으로
operational site(ramp area) 항공기 운용 장소(램프 지역)
preventive maintenance 예방 정비
provide 제공하다
is (be) regarded as ~ ~로 간주(취급)하다
repeat 반복하다
requirements 요목, 요건
specific 특정한
structural parts 기체구조 부품(부재)
task card 작업 카드
various 다양한

Translation

정시 점검은 항공기의 여러 계통 및 기체 구조부에 대한 예방 정비이자 감항성을 확인하는 것이다. 이 점검은 다양한 종류의 작업 카드를 사용하여 수행하며, 비행 시간, 비행 횟수, 또는 날/달 또는 년수로 정해진 주기에 따라 반복 수행한다.

제작사에서는 정기 점검 요목들을 개별 항목으로 제공하나, 항공사는 편의에 따라 "A", "C", "D" 점검 등의 형태나 특정한 그룹으로 묶을 수 있다.

통상 "A" 점검은 운항 정비로 취급되어 운용 현장(램프 등)에서 수행할 수 있으나, "C", "D" 점검은 중정비로서 격납고와 같은 정비시설 안에서 수행한다.

4.1 Practice Quiz

Answer Keys p. 254

01 Which maintenance is not performed routinely for aircraft operation?

a. Regular Maintenance
b. Project Maintenance
c. Unscheduled Maintenance
d. Scheduled Maintenance

02 Which maintenance is performed unexpectedly to check for abnormalities in the aircraft when the aircraft experienced an abnormal condition?

a. Regular Maintenance
b. Project Maintenance
c. Unscheduled Maintenance
d. Scheduled Maintenance

03 Which is the check performed at the departure base and intermediate base whenever the flight changes?

a. Pre/Post Flight Check
b. Transit Check
c. Weekly Check
d. Periodic Check

04 Which is the preventive maintenance and airworthiness checks for various systems and structural parts of the aircraft.

a. Pre/Post Flight Check
b. Transit Check
c. Weekly Check
d. Periodic Check

05 Which check is performed once between the last flight of the day and the first flight of the next day?

a. Pre/Post Flight Check
b. Transit Check
c. Weekly Check
d. Periodic Check

06 다음 중 아래 문장의 내용상 ( )에 적합한 단어는?

These checks are performed by using various kind of task cards and are repeated at the (   ) determined by flight hours, flight cycles, or days/months or years.

a. calendar time
b. gaps
c. intervals
d. distances

PART 4.2 Maintenance Program

4.2.1 Structure of Maintenance Program

The aircraft maintenance program consists of various maintenance requirements. The main parts of this program are System & Powerplant Maintenance Program, Structural Maintenance Program, and Zonal Inspection Program.

Words & Phrase

consists of ~ ~으로 구성되다
powerplant 동력장치(엔진 및 관련 시스템)
requirements 요구사항, 요목
zonal 구역별, 구역의
introduce ~ ~을 소개한다
system 계통
various types of 다양한 종류의

Translation

항공기 정비 프로그램은 여러 가지의 정비 요목들로 구성되어 있다. 이 프로그램의 주요 부분은 시스템/동력장치 정비 프로그램, 기체 정비 프로그램 그리고 구역별 검사 프로그램이다.

1) Systems & Powerplant Maintenance Program

This program specifies the scheduled maintenance tasks for all of the aircraft systems & powerplant and use following types of task.

ATA Chapters: 12, 20, 21~49, 51~57, 70~80

Task Type: SVC, LUV, VCK, GVI, DET, OPC, FNC, RST, DIS.

Words & Phrase

scheduled maintenance tasks 계획 정비 작업
specify ~ ~을 열거하다 (~을 명시하다)

Translation

이 프로그램은 항공기의 모든 시스템과 동력장치에 대한 계획 정비 작업 요목들을 명시하고 있으며 다음과 같은 종류의 작업을 이용한다.

해당 ATA Chapters: 12, 20, 21~49, 51~57, 70~80

작업 종류: 보급, 윤활, 육안 점검, 일반 육안 검사, 상세 검사, 작동 점검, 기능 점검, 기능 회복, 부품 교환.

SYSTEMS AND POWERPLANT MAINTENANCE PROGRAM

MPD ITEM NUMBER	AMM REFERENCE	CAT	TASK	INTERVAL		ZONE	ACCESS	APPLICABILITY		MAN-HOURS	TASK DESCRIPTION
				THRESH	REPEAT			APL	ENG		
12-027-04	12-21-05-100 12-21-05-640	6 9	SVC	8 YR	8 YR	200	NOTE	ALL	ALL	0.40	Service by non solvent cleaning and lubricating the aileron cables in pressurized areas. Note: CRES cables should not be lubricated. ACCESS NOTE: Gain access as required.

[그림 4-2] 계통/동력장치 정비 프로그램 (사례)

2) Structural Maintenance Program

This program specifies the scheduled maintenance tasks for the aircraft structures and use following types of task.

ATA Chapters: 32, 52, 53, 54, 55, 57.

Task Type: GVI, DET, SDI.

Words & Phrase

scheduled maintenance tasks 계획 정비 작업
specify ~ ~을 열거하다 (~을 명시하다)

Translation

이 프로그램은 항공기 기체 구조 부분에 대한 계획 정비 작업 요목들을 명시하고 있으며 다음과 같은 종류의 작업을 이용한다.

해당 ATA Chapters: 32, 52, 53, 54, 55, 57.

작업 종류: 일반 육안 검사, 상세 검사, 특별 상세 검사.

STRUCTURAL MAINTENANCE PROGRAM

MPD ITEM NUMBER	AMM REFERENCE	PGM	ZONE	ACCESS	INTERVAL		APPLICABILITY		MAN-HOURS	TASK DESCRIPTION
					THRESH	REPEAT	APL	ENG		
53-400-00	53-05-03-211	S	112	801 NOTE	8 YR 9600 FC NOTE	6 YR 7200 FC NOTE	FRTR	ALL	1.00	*ATA 53: FUSELAGE* *INTERNAL - DETAILED*: Area Forward of Nose Wheel Well Detailed Inspection of the nose cargo door cutout INTERVAL NOTE: Whichever comes first. After 8YR/9600FC a repeat interval of 6YR/7200FC is applicable. ACCESS NOTE: For access remove or displace all systems, equipment and interior furnishings as necessary to accomplish visual inspection.

[그림 4-3] 기체 정비 프로그램 (사례)

3) Zonal Inspection Program

This program specifies the inspection of the general condition and security of accessible systems and structural items contained in defined zone of the aircraft. The program also includes a condition check of the components of the electrical wiring connection system and uses the following types of task.

ATA Chapters: 32, 52, 53, 54, 55, 57, 71.

Task Type: VCK, GVI.

Words & Phrase

accessible 접근할 수 있는 (접근이 가능한)
components 구성품
contained in~ ~에 포함된
defined zone 한정된 구역
electrical wiring connection system 전기배선 연결 체계
include ~을 포함하다
security 고정(상태)
structural items 기체구조 구성품(부재)

Translation

이 프로그램은 항공기의 특정 구역 안에 포함된 접근 가능한 시스템 및 기체 부품에 대한 일반적인 상태와 고정 상태의 점검을 명시하고 있다. 이 프로그램은 전기배선 연결 시스템의 구성품에 대한 상태 점검도 포함하며 다음과 같은 종류의 작업을 이용한다.

해당 ATA Chapters: 32, 52, 53, 54, 55, 57, 71.

작업 종류: 육안 점검, 일반 육안 검사

ZONAL INSPECTION PROGRAM

MPD ITEM NUMBER	AMM REFERENCE	ZONE	ACCESS	INTERVAL		APPLICABILITY		MAN-HOURS	TASK DESCRIPTION
				THRESH	REPEAT	APL	ENG		
									ATA 71: POWER PLANT
71-802-01	05-41-04-212	410 420 430 440		4000 FH	4000 FH	ALL	ALL	2.00	*EXTERNAL - ZONAL (GV)*: POWERPLANT ENGINE NO. 1, 2, 3, 4 General Visual Inspection of the powerplant engine no. 1, 2, 3, 4.
71-804-01	05-41-04-212	411 421 431 441	NOTE	2 YR 2400 FC NOTE	2 YR 2400 FC NOTE	ALL	ALL	1.40	*EXTERNAL - ZONAL (GV)*: NOSE COWL, ENGINE NO. 1, 2, 3, 4 General Visual Inspection of the nose cowl, engine no. 1, 2, 3, 4. (EZAP) INTERVAL NOTE: Whichever comes first. ACCESS NOTE: Open fan cowls.

[그림 4-4] 구역별 검사 프로그램 (사례)

4.2.2 Task Types of Maintenance Program

1) SVC/LUB (SERVICE/LUBRICATION)

The term "Service" or "Lubrication" implies that a component or system should be checked and serviced with fuel, oil, grease, water, oxygen, etc., to a level or condition specified by the appropriate Maintenance Manual procedure. "SERVICE" may also be used to indicate that filter cleaning or replacement is recommended.

Words & Phrase			
	appropriate 관련된, 적절한	**cleaning** 세척	**condition** 상태
	imply 의미하다	**indicate** 나타내다	**level** 수준
	recommend 권고하다	**replacement** 교환(교체)	**specify** 명시하다

Translation

"서비스" 또는 "윤활" 이라는 용어는 장비품이나 시스템을 점검하고 적절한 정비 매뉴얼의 절차에 명시한 수준이나 조건에 맞게 연료, 오일, 그리스, 물, 산소 등을 보충하는 것을 의미한다. "서비스" 라는 용어는 필터의 청소나 교체를 권고하는 것으로도 사용할 수 있다.

2) VCK (VISUAL CHECK)

A visual failure finding task through observation to determine that an item is fulfilling its intended purpose. This task does not require quantitative tolerances.

Words & Phrase		
	determine 알아내다, 결정하다	**fulfill** 충족하다
	intended purpose 의도한 목적	**observation** 관찰
	quantitative 수량적인, 수치로 주어지는	**tolerances** 허용치
	visual 육안의	

Translation

육안 점검은 점검의 대상이 의도한 목적을 충족하고 있는지 알아내는 것으로, 관찰을 통해 육안으로 결함을 찾는 작업이다. 이 작업은 수치로 주어지는 허용치가 필요 없다.

3) GVI (GENERAL VISUAL INSPECTION)

A visual examination of an interior or exterior area, installation or assembly to detect obvious damage, failure or irregularity. This level of inspection is made from within touching distance unless otherwise specified. A mirror may be necessary to ensure visual access to all surfaces in the inspection area.

This level of inspection is made under normally available lighting conditions such as daylight, hangar lighting, flashlight or drop-light and may require removal or opening of access panels or doors. Stands, ladders or platforms may be required to gain proximity to the area being checked.

Words & Phrase

available 이용할 수 있는, 사용할 수 있는
detect 감지하다, 발견하다
ensure ~을 보장하다, 반드시 ~하게 하다
exterior 외부
flashlight 손전등
interior 내부
ladder 사다리
mirror 거울
platforms 작업용 받침대
to gain proximity to ~ ~에 가깝도록, ~에 접근하기 위해
touching distance 손이 닿는 거리
daylight 일광 (햇빛)
drop-light 매달린(늘어진) 등불
examination 검사
failure 고장
hangar lighting 격납고 조명
irregularity 비균일성, 불규칙성, 고르지 않음
lighting condition 조명 상태
obvious 분명한, 명백한
stands 지지대, 작업대
unless otherwise specified 별도로 명시되지 않으면

Translation

명백한 손상, 고장 또는 불규칙성을 감지하기 위한 내 · 외부, 장착 품이나 조립품에 대한 자세한 육안 검사이다. 이 수준의 검사는 달리 명시하지 않아도 손이 닿을 수 있는 거리 내에서 수행한다. 검사 부위의 모든 표면을 육안으로 보기 위해 거울이 필요할 수도 있다.

이 수준의 검사는 태양광, 격납고 조명, 손전등 또는 낙하 등과 같은 일반적인 조명 아래에서 수행하며 액세스 패널 또는 도어를 제거하거나 여는 것이 필요할 수 있다. 검사 대상 부위에 접근하기 위해 스탠드, 사다리 또는 플랫폼이 필요할 수 있다.

4) DET (DETAILED INSPECTION)

An intensive visual examination of a specific item, installation, or assembly to detect damage, failure, or irregularity. Available lighting is normally supplemented with a direct source of good lighting at an intensity deemed appropriate. Inspection aids such as mirrors, magnifying lenses, etc. may be necessary. Surface cleaning and elaborate access procedures may be required.

Words & Phrase

aids 보조기구 **available** 이용할 수 있는, 사용할 수 있는
deem ~으로 생각하다 **detailed** 상세한 **detect** 찾아내다, 탐지하다
elaborate 정교한 **examination** 검사 **inspection aids** 검사 보조기구
intensity 강도, 조명도 **intensive** 강력한, 집중적인 **lighting** 조명
magnifying lenses 확대 렌즈 (확대경) **mirrors** 거울
specific 특정한 **is supplemented with** ~으로 보충(보완)하다

Translation

손상, 고장 또는 불규칙성을 감지하기 위한 특정 품목, 장착물 또는 조립품에 대한 집중적인 육안 검사이다. 사용할 수 있는 조명 외에 통상 적절한 강도의 직접 조명을 추가로 사용한다. 거울, 돋보기 등 검사 보조 도구가 필요할 수 있다. 표면 청소 및 정교한 접근 절차가 필요할 수도 있다.

5) SDI (SPECIAL DETAILED INSPECTION)

An intensive examination of a specific item(s), installation or assembly to detect damage, failure or irregularity. The examination is likely to make extensive use of specialized inspection techniques and/or equipment. Intricate cleaning and substantial access or disassembly procedures may be required.

Words & Phrase

disassembly 분해 **examination** 검사 **extensive** 광범위한
intensive 강한, 집중적인 **intricate** 복잡한 **is likely to~** ~할 가능성이 크다
specialized 전문화된 **substantial** 상당한 **technique** 기법, 기술

Translation

손상, 고장 또는 불규칙성을 감지하기 위한 특정 품목, 장착물 또는 조립품에 대한 집중적인 검사이다. 이 검사에서는 전문화된 검사 기술 및/또는 장비를 광범위하게 사용할 가능성이 크다. 복잡한 세척 작업과 상당한 접근 또는 분해 절차가 필요할 수도 있다.

6) OPC (OPERATIONAL CHECK)

A test used to determine that a system or component or any function thereof is operating normally.

Words & Phrase		
	determine 결정(확정)하다, 알아내다 **~ thereof** 그것의 ~	**function** 기능

Translation

시스템이나 구성 요소 또는 그것의 기능이 정상적으로 작동하는지 확정하는 데 사용되는 테스트이다.

7) FNC (FUNCTIONAL CHECK)

A detailed examination in which a complete system, sub-system or component is checked to determine if operating parameters are within limits of movement, rate of flow, temperature, pressure, revolutions per minute, degrees of travel, etc., as prescribed in the manufacturer/vendor Maintenance Manual.

Words & Phrase		
	as prescribed in ~ ~에 미리 정해진 **detailed examination** 정밀 검사 **parameters** 변수(기능의 상태를 보여주는 수치) **rate of flow** 유량, 유출 속도 **temperature** 온도	**degrees of travel** 작동(움직임) 각도 **movement** 운동, 활동 **pressure** 압력 **revolutions per minute** 1분당 회전수

Translation

전체 시스템, 하위 시스템 또는 구성 요소의 작동 변수가 제작사/생산자의 정비 매뉴얼에 규정되어 있는 운동, 유량, 온도, 압력, 분당 회전수, 움직임 각도 등에 대한 허용치 내에 있는지 확인하기 위해 하는 상세 검사이다.

8) RST (RESTORATION)

The work necessary to return the item to a specific standard. Restoration may vary from cleaning or replacement of single parts up to a complete overhaul.

Words & Phrase

overhaul 분해 검사
specific 특정한
restoration (기능) 회복
standard 기준
return 되돌리다
vary 변하다, 바뀌다, 다르다

Translation

대상 품목을 특정한 기준(상태)으로 되돌리는 데 필요한 작업이다. 기능 회복은 단일 부품을 세척하거나 교체하는 것부터 완전 오바홀(분해수리)하는 것에 이르기까지 다양하다.

9) DIS (DISCARD)

This action is a permanent removal from service of an item or assembly at a specified life limit.

Words & Phrase

discard 폐기하다, 버리다, 폐기
specified life limit 명시된 수명 한계
permanent removal 영구 제거

Translation

이 작업은 특정한 수명 한계에 도달하면 어떤 품목이나 조립품을 영구적으로 사용하지 못하게 하는 것이다.

4.2 Practice Quiz

Answer Keys p. 254

01 Which maintenance program does include a condition check of the components of the EWIS (electrical wiring connection system)?

a. System & Powerplant Maintenance Program
b. Structural Maintenance Program
c. Zonal Inspection Program

02 Which is the task that asks check and service with fuel, oil, grease, water, to a level or condition specified by the appropriate Maintenance Manual?

a. Visual Check
b. Service
c. General Visual Inspection
d. Restoration

03 Which task is the intensive visual examination of a specific item, installation, or assembly to detect damage, failure, or irregularity.

a. Visual Check
b. General Visual Inspection
c. Detailed Inspection
d. Special Detailed Inspection

04 Which is the task to check and determine if operating parameters are within limits prescribed in the manufacturer/vendor Maintenance Manual.

a. General Visual Inspection
b. Detailed Inspection
c. Operational Check
d. Functional Check

05 Which task is for permanent removal from service of an item at a specified life limit?

a. Service
b. Discard
c. Restoration
d. Lubrication

06 Which task is not belong to visual inspection?

a. Visual Check
b. General Visual Inspection
c. Detailed Inspection
d. Special Detailed Inspection

07 다음 Task 중에서 작업을 위해 별도의 측정기기가 필요한 것은?

a. Operational Check
b. Functional Check
c. Restoration
d. Special Detailed Inspection

08 다음은 어떤 Task에 관해서 설명하는 것인가?

> "Available lighting is normally supplemented with a direct source of good lighting at an intensity deemed appropriate."

a. Visual Check
b. General Visual Inspection
c. Detailed Inspection
d. Special Detailed Inspection

PART 4.3 Maintenance Task Card

4.3.1 Technical English in Task Card

1) 작업 절차에는 동사가 문장의 맨 앞에 나오는 명령문을 사용한다.

Work procedures use imperative sentences where the verb comes first of the sentence.

Words & Phrase		
	access panel 점검창 (점검을 위해 장탈·착 할 수 있는 패널)	
	~comes first ~이 먼저 온다, ~이 앞서다	**imperative** 명령법의
	procedure 절차	**verb** 동사
	work 작업	**where** 거기에서는

Example

Open the forward cargo compartment door.
전방 화물 실의 문을 열어라.

Remove the access panel.
악세스 패널을 장탈하라.

2) 작업 대상의 조건(형태, 상태)에 따라 작업의 절차나 내용이 달라질 때는, (If~)로 시작하는 가정법 문장이 전체 문장의 맨 앞에 온다.

If the procedure or content of the work changes depending on the configuration or state of the object, the sentence beginning with (If~) comes first of the entire sentence.

Words & Phrase			
	begin with ~로 시작하다.	**configuration** 형태(형상)	**depending on~** ~에 따라서
	lockwire 고정 와이어	**object** 사물	**state** 상태

Example

If there is a lockwire on the coupling, make sure it is not loose.
(configuration of the object) (contents of work to do)
커플링에 록 와이어가 되어 있다면, 그것이 느슨하지 않은지 확인하라.

If there is no crack on the fitting, perform step (d).
(state of the object) (procedure to follow)
피팅에 균열이 없다면 (d) 항의 절차를 수행하라.

3) 대부분은 능동태의 문장을 사용한다.

In most cases, sentences in the active voice are used.

Words & Phrase			
	active voice 능동태	**altitude** 고도	**cabin** 객실
	in most cases 대부분은	**passive voice** 수동태	**pressure** 압력

Example

능동태 The cabin pressure controller (subject/주어) controls (verb/동사) the cabin altitude (object/목적어).

객실 압력 조절기는 객실 고도를 조절한다.

수동태 The cabin altitude (object/목적어) is controlled by (verb/수동태) the cabin pressure controller (subject/주어).

객실 고도는 객실 압력 조절기에 의해 조절된다.

4) 작업 대상을 꾸미는 문장이나 작업 내용을 설명하는 문장은 "which"나 "that"으로 시작한다.

Sentences that describe the object of work or describe the contents of the work begin with "which" or "that"

Words & Phrase			
	access panel 점검창	**contents** 내용	**describe** 꾸미다, 서술하다.
	fasteners 패스너(rivet, bolt 같이 두 개 이상의 물체를 고정 또는 결합하는 부품)		
	make sure that~ ~을 확인하다		**release** 풀다, 풀어주다.

Example

Release the fasteners which attach the access panel to the filter housing.

(which 뒤의 문장은 which 앞의 "fasteners"를 꾸미는 말임)

필터 하우징의 악세스 패널을 부착하고 있는 패스너들을 풀어 주어라.

Make sure that all spring clips are installed and not damaged.

(that 이하 문장이 작업(확인)할 내용임)

모든 스프링 클립들이 장착되어 있고 손상되지 않았는지 확인하라.

5) 주어가 길면 가주어(it)나 유도 부사(there) 등을 사용한다.

If the subject is long, use dummy subjects (it) or inductive adverb (there)

Words & Phrase			
	dummy 가짜의	**inductive adverb** 유도 부사	**real** 진짜의

Example

It is not required ———— to remove the access panel.
(dummy subject/가주어) (real subject/ 진주어)
악세스 패널을 장탈하는 것은 필요하지 않다.

There are ———— three different procedures that you can use.
(inductive adverb/유도 부사) (real subject/진주어)
당신이 이용할 수 있는 3개의 서로 다른 절차가 있다.

6) 특정한 전치사가 기술 영어 문장에 빈번하게 사용된다.

Specific prepositions are frequently used in the technical English sentences.

Example to ~

~(으)로, ~에(게), ~까지, ~하기 위해, ~하는 것

~으로 Put ATC select switch to the NO. 2 position.
ATC 선택 스위치를 NO. 2 위치로 놓아라.

~에 The test is applicable to the left and the right ATC system.
Test는 좌·우측 ATC 시스템에 적용된다.

~까지 The cables carry the high voltage DC from the exciters to the igniters.
케이블은 고전압의 직류 전기를 익싸이터에서 이그나이터까지 전달한다.

~하려면 You will need 3 persons to complete this task.
이 작업을 완료하려면 세 사람이 필요하다.

~하는 것 It is not required to remove the access panel.
악세스 패널을 장탈하는 것은 필요하지 않다.

Example for ~

~동안, ~의(~에 대한), ~가 있는지

~동안 Let the potting compound dry for one hour.
팟팅 컴파운드를 1시간 동안 건조하라.

~에 대한 Obey the removal procedure for the components.
컴포넌트에 대한(의) 장탈 절차를 준수하라.

~가 있는지 Examine the casting for corrosion.
캐스팅에 부식이 있는지 자세히 검사하라.

Example with ~

~과(~에), ~을 사용하여, ~을 이용하여, ~한 상태로, ~가 있는

~과/~에 Align the mark with longitudinal axis.
표시를 길이 방향의 축과(에) 정렬하라(일직선으로 하다).

~을 사용하여 Clean the filter housing with a cotton wiper.
필터 하우징을 면 걸레를 사용하여 닦아라.

~을 이용하여 The center lower display unit is installed with the latch mechanism at the top.
중앙 하단의 디스플레이는 위쪽에 있는 래치(걸쇠)를 이용하여 장착된다.

~가 있는 This procedure is applicable to the airplane with GPS.
이 절차는 GPS가 있는 항공기에 해당한다.

4.3.2 Typical Sentences of Task Card

1) Servicing

If you see bright silver mark, service IDG with Mobile Jet II MIL-PRF-23699, Type II.

If the oil level is below the top thread of the oil hole or you can not see the oil level, carefully add Mobil Jet II oil into the oil fill port.

Words & Phrase			
	add 보충하다	**bright** 밝은	**IDG** Integrated Drive Generator
	oil fill port 오일 주입구	**service** 보급하다, 보충하다	**thread** 나사산

Translation

밝은 Sliver Mark가 보이면 IDG에 오일(Mobile Jet II, MIL-PRF-23699,Type II)을 보충해 주어라.

오일의 액면이 오일 주입구의 나사산 맨 위쪽보다 낮거나 오일의 액면을 볼 수 없으면, 오일 주입구를 통해 오일(Mobil Jet II)을 주의하여 보충하라.

Sentence Analysis

If you see bright silver mark, (작업 조건) / service IDG (작업 종류/대상) / with Mobile Jet II MIL-PRF-23699,Type II. (작업 수단)

If the oil level is below the top thread of the oil hole or you can not see the oil level, (작업 조건) / carefully add Mobil Jet II oil (작업 종류/수단) / into the oil fill port. (작업 대상/위치)

2) Lubrication

Lubricate the jackscrew ballnut and the stabilizer trim jackscrew with BMS 3-33, D00633 grease. (Figure 1)

Lubricate the No.4 and No.5 flap ballscrew nut, universal joint and jimbal cap with flaps in the 30-unit detent position, using a non-pneumatic hand-held grease gun and BMS 3-33 grease. (Figure 1)

Words & Phrase			
	detent 멈춤쇠	**grease gun** 그리스 주입기	**hand-held** 휴대용
	lubricate 윤활하다	**non-pneumatic** 비공압식	
	stabilizer trim jackscrew 스태빌라이저 조절 잭 스크루		

Translation

잭 스크루 볼너트와 스태빌라이저 트림 잭 스크루를 그리스(BMS 3-33, D00633)를 사용하여 윤활하라.

플랩을 30-unit 위치에 놓은 상태에서, 비공압식 그리스 주입기와 그리스(BMS 3-33)를 사용하여 No. 4 및 No. 5 플랩 볼 스크루 너트, 유니버설 조인트 및 짐벌을 윤활하라.

Sentence Analysis

Lubricate (작업 종류) / the jackscrew ballnut and the stabilizer trim jackscrew (작업 대상) / with BMS 3-33, D00633 grease (작업 수단). (Figure 1) (참조 그림)

Lubricate (작업 종류) / the No.4 and No.5 flap ballscrew nut, universal joint and jimbal cap (작업 대상) / with flaps in the 30-unit detent position, (작업을 위한 항공기 상태/조건) / using a non-pneumatic hand-held grease gun and BMS 3-33 grease (작업 수단). (Figure 1) (참조 그림)

3) Visual Check (VCK)

Check the CREW OXY LOW advisory message or the CREW OXY REFILL message is not shown on the EICAS status page.
Check that the portable oxygen cylinders are correctly installed to the wall-mounted brackets or storage compartment.

Words & Phrase		
	advisory 주의, 알림	**correctly** 적절한, 정확한
	portable oxygen cylinders 휴대용 산소통	**status** 상태
	storage compartment 저장실 (수납공간)	**wall-mounted bracket** 벽에 장착된 걸개 (벽걸이)

Translation

CREW OXY LOW 알림 메시지나 CREW OXY REFILL 메시지가 EICAS 상태 페이지 상에 나타나지 않는지 점검하라.

휴대용 산소통이 벽걸이나 수납 공간에 올바르게 정착되어 있는지 점검하라.

Sentence Analysis

Check / the CREW OXY LOW advisory message or the CREW OXY REFILL message / is not shown / on the EICAS status page.

(작업 종류) / (작업 대상) / (점검 내용) / (점검 위치)

Check that / the portable oxygen cylinders / are correctly installed to the wall-mounted brackets or storage compartment.

(작업 종류) / (작업 대상) / (점검 내용)

4) General Visual Inspection (GVI)

Do an inspection of the inner vane platform for cracks.

Perform a General Visual Inspection of nose landing gear from the ground.

Words & Phrase		
	crack 균열 **inner** 안쪽의	**general visual inspection** 일반 육안 검사

Translation

이너 베인 플랫폼에 균열이 있는지 검사하라.

지상에서 노즈 랜딩 기어에 대한 일반 육안 검사를 수행하라.

Sentence Analysis

Do an inspection / of the inner vane platform / for cracks.
(작업 종류) (작업 대상) (점검 내용)

Perform a General Visual Inspection / of nose landing gear / from the ground.
(작업 종류) (작업 대상) (작업 위치)

5) Detailed Inspection (DET)

If bright metal particles or larger, non-metallic, black colored, fiber-like pieces are present in considerable quantity on the filter element, in the filter cavity, or in the oil drained from the IDG, do these steps.
Perform a Detailed Inspection of the cable runs for incorrect routing, kinks in the cable, or other damage.

Words & Phrase

black colored 검은 색의	**cable runs** 케이블 배선
cavity 동공(빈 부분), 구멍	**considerable quantity** 상당한 양
detailed inspection 상세 육안 검사	**drain** ~을 빼다
fiber-like 섬유질의	**incorrect routing** 잘못된 경로
kinks 뒤틀림	**metal particles** 금속 입자
non-metallic 비금속의	

Translation

밝은 빛의 금속 입자 또는 조금 더 큰 검은 색의 섬유질 조각이 상당한 양으로 필터 표면, 필터 구멍 안이나 IDG에서 빼낸 오일에 존재할 때는 이 작업(단계)들을 수행하라.
케이블 배선에 잘못된 경로나 뒤틀린 곳이나 그 밖에 손상이 있는지 상세한 검사를 수행하라.

Sentence Analysis

If bright metal particles or larger, non-metallic, black colored, fiber-like pieces are present in considerable quantity (작업 조건 / 결함 내용) / on the filter element, in the filter cavity, or in the oil drained from the IDG, (결함 발견 위치) / do these steps. (조치 내용)

Perform a Detailed Inspection (작업 종류) / of the cable runs (작업 대상) / for incorrect routing, kinks in the cable, or other damage. (점검 내용)

6) Operational Check

Push and hold the TEST switch on the voice recorder's panel (P461) for a minimum of 5 seconds.
Supply electrical power (AMM 24-22-00/201), Check the EICAS Status message ALTN VENT RELAY does not show on the lower EICAS display.

Words & Phrase	
electrical power 전력	**hold** ~하고 있다
status message 상태를 알려주는 메시지	**supply** 공급하다.
voice recorder 음성 녹음기	

Translation

voice recorder' s panel(P461)에서 테스트 스위치를 최소 5초 동안 누르고 있어라.
전력을 공급하라(AMM 24-22-00/201 참고). EICAS 상태 메시지 "ALTN VENT RELAY"가 아래쪽 EICAS 화면 위에 나타나지 않는지 점검하라.

Sentence Analysis

Push and hold / the TEST switch / on the voice recorder's panel (P461) /
(작동 구분) (작동 대상) (작동 대상의 위치)
for a minimum of 5 seconds.
(작동 시간)

Supply electrical power (AMM 24-22-00/201), / Check / the EICAS Status
(작동 조건) (작업 내용) (점검 대상)
message ALTN VENT RELAY / does not show / on the lower EICAS display.
(점검 대상) (점검 내용) (점검 위치)

7) Functional Check

Check the tire pressure value shown for each tire on the TBMS (Tire Pressure and Brake Monitoring System) is within +/- 10 psig of the tire pressure value measured with the tire pressure gage at the tire.
If the valve opens at more than 9.35 psi (64.5 kPa), make sure there are no loose fittings on the remote ambient sense tubing or the test equipment tubing.

Words & Phrase		
	ambient 주위(주변)의	**measured with** ~으로(을 사용하여) 측정한
	pressure gage 압력 게이지	**remote** 먼, 멀리 떨어진
	be(is) within xxx of ~ ~의 xxx 내 (xxx 범위 안)	

Translation

타이어 압력 및 브레이크 모니터링 시스템에서 각각의 타이어 별로 표시된 타이어 압력 값이 타이어 압력 게이지를 사용하여 측정한 값의 +/- 10 psig 범위 안에 있는지 점검하라.
만일 밸브가 9.35 psi보다 높은 압력에서 열린다면, 먼 쪽에 있는 주변 압력 감지 튜브나 테스트 장비 튜브에 헐거워진 피팅이 없는지 확인하라.

Sentence Analysis

Check (작업 종류) / the tire pressure value shown for each tire on the TBMS (Tire Pressure and Brake Monitoring System) (점검 대상) / is within +/- 10 psig of the tire pressure value measured with the tire pressure gage at the tire. (점검 내용)

If the valve opens at more than 9.35 psi (64.5 kPa), (작업 조건) / make sure there are no loose fittings (점검 내용) / on the remote ambient sense tubing or the test equipment tubing. (점검 위치)

8) Restoration

Use a soft bristle brush to clean the inlet screen with soap water.
Install the current NDB (Navigation database) into the LH FMC (Flight Management Computer) per AMM 34-62-00/201.

Words & Phrase			
	bristle 털	**current** 현재(행)의	**Flight Management**
	Computer 비행 관리 컴퓨터	**inlet** 입구	**install** ~을 설치하다, ~을 장착하다
	navigation database 항법용 데이터베이스		**soap water** 비눗물

Translation

입구 스크린을 비눗물로 씻으려면 부드러운 털 브러쉬를 사용하라.
현행(최신판)의 항법용 데이터베이스를 AMM 34-62-00/201의 절차에 따라서 좌측 FMC (비행 관리 컴퓨터)에 설치하라.

Sentence Analysis

Use a soft bristle brush (작업 수단) / to clean the inlet screen (작업 목적) / with soap water (사용 자재).

Install (작업 종류) / the current NDB (Navigation database) (작업 내용) / into the LH FMC (Flight Management Computer) (작업 대상) / per AMM 34-62-00/201 (작업 근거).

9) Discard

Remove and discard / the packing and the oil filter element.

Words & Phrase		
	discard 폐기하다	**filter element** 필터 소자 (여과기 몸체)

Translation

패킹과 오일 필터 엘리먼트를 장탈해서 폐기하라.

Sentence Analysis

Remove and discard / the packing and the oil filter element.

(작업 종류) (작업 내용)

4.3.3 Typical Format of Task Card

The following is a sample of aircraft maintenance task card. The format of the card may differ slightly depending on the model or air carrier, but the content is taken from the manufacturer's AMM, so there is no significant difference.

Words & Phrase		
air carrier 항공회사(운송회사)	**depending on** ~에 따라	**differ slightly** 약간 다르다.
format 구성 방식	**significant** 상당한, 중대한	

Translation

다음은 항공기 정비 작업 카드의 사례이다. 카드의 형식은 항공기 모델이나 항공사에 따라 다소 다를 수 있으나, 그 내용은 항공기 제작사의 AMM에서 가져온 것이므로 큰 차이가 없다.

<table>
<tr><td colspan="2">❶ AIRLINE CARD NO</td><td colspan="3">❹ TITLE</td><td colspan="2">❷ MANUFACTURER CARD NO.</td></tr>
<tr><td colspan="2"></td><td colspan="3" rowspan="3">E/E COOLING SUPPLY FAN FILTER</td><td colspan="2">21-040-XX-XX</td></tr>
<tr><td>❺ DATE</td><td>❻ TASK</td><td colspan="2">❸ RELATED CARD</td></tr>
<tr><td></td><td>REPLACE</td><td colspan="2"></td></tr>
<tr><td rowspan="2">❼ TAIL NUMBER</td><td rowspan="2">❽ WORK AREA</td><td rowspan="2">⓫ VERSION</td><td rowspan="2">⓬ THRESHOLD</td><td rowspan="2">⓭ REPEAT</td><td colspan="2">APPLICABILITY</td></tr>
<tr><td>⓮ AIRPLANE</td><td>⓯ ENGINE</td></tr>
<tr><td></td><td>FWD CARGO</td><td>1.1</td><td>7500 FH</td><td>7500 FH</td><td>ALL</td><td>ALL</td></tr>
<tr><td>❾ STATION</td><td>❿ SKILL</td><td colspan="3">⓰ ACCESS</td><td colspan="2">⓱ ZONE</td></tr>
<tr><td></td><td>AIRPL</td><td colspan="3"></td><td colspan="2">118</td></tr>
<tr><td colspan="7">⓲
Replace the E/E cooling supply fan filter.
A. References
Reference — Title
AMM 24-22-00-XXX-XXX — Remove Electrical Power (P/B 201)
AMM 25-52-16-XXX-XXX — Forward Cargo Compartment Forward Bulkhead Liner-Removal (P/B 401)
AMM 25-52-16-XXX-XXX — Forward Cargo Compartment Forward Bulkhead Liner-Installation (P/B 401)
AMM 52-31-00-XXX-XXX — Open the Cargo Door (P/B 201)
AMM 52-31-00-XXX-XXX — Open the Cargo Door (P/B 201)</td></tr>
<tr><td colspan="2">⓳ EFFECTIVITY</td><td>⓴ SOURCE</td><td colspan="4" rowspan="2">E/E COOLING SUPPLY FAN FILTER ❹
㉑ Page 1 of 5
㉒ Jun 15/2024</td></tr>
<tr><td colspan="2"></td><td>MRB</td></tr>
</table>

[그림 4-5] 작업 카드의 형식 (사례)

1) Column of Task Card

❶ AIRLINE CARD NO	항공사가 부여한 Task Card 번호 (예: 737-21-XXX-XX-01)
❷ MANUFACTURER NO	제작사가 부여한 Task Card 번호 (21-040-XX-01)
❸ RELATED CARD NO	본 Task Card와 관련된 다른 Task Card 번호 (없음-공란)
❹ TITLE	Task Card의 제목 (E/E COOLING SUPPLY FAN FILTER)
❺ DATE	작업 수행일
❻ TASK	Task의 종류 (REPLACE = DISCARD)
❼ TAIL NUMBER	항공기의 등록번호 (HLXXXX)
❽ WORK AREA	작업 부위 (FWD CARGO = FWD CARGO COMPARTMENT)
❾ STATION	작업을 수행한 장소 (3 LETTER - ICN / 인천)
❿ SKILL	작업자 특기 (기체: AIRPL, 엔진: ENGIN, 전기: ELEC 등)
⓫ VERSION	Task Card의 형식 (발행처의 개정 관리번호: 1.1, 1.2, 1.3, …)
⓬ THRESHOLD	Task Card의 초도 수행 시기 (7500 FH: 누적 비행시간 7500시간)
⓭ REPEAT	반복 점검 주기 (7500 FH: 비행시간 7,500시간 간격으로 반복 수행)
⓮ APPLICABILITY	Task Card 적용 항공기 Type
- AILPLANE	(ALL - 항공사 보유 특정 기종 항공기 전체)
⓯ APPLICABILITY	Task Card 적용 엔진 Type
- ENGINE	(ALL - 항공사 보유 특정 기종 항공기의 전체 엔진)
⓰ ACCESS	점검을 위해 장탈해야 하는 점검창 번호 (예: 118AL, …)
⓱ ZONE	작업 구역 (118: Forward Cargo Compartment)
⓲ Task summary & procedures (작업 개요 및 작업 절차)	
⓳ EFFECTIVITY	Task Card가 적용되는 항공기의 번호 (항공사 특정 기종의 항공기들의 Manual 적용 코드 번호/ 예: ABC002 등)
⓴ SOURCE	Task Card 제정 근거 (MRB: FAA 발행 Maintenance Review Board Report에 포함된 필수 점검 항목)
㉑ PAGE	Task Card의 쪽 번호 (Page 1 of 5: 전체 5쪽 중 1쪽)
㉒ REVISION DATE	Task Card의 발행일.

4.3.4 Contents of Task Card

A. References

※ 본 Task 수행 시 수행하게 되는 작업 사항 및 관련 정비 도서의 근거와 제목을 소개

Reference	Title
AMM 24-22-00-XXX-XXX	Remove Electrical Power (P/B 201)
AMM 25-52-16-XXX-XXX	Forward Cargo Compartment Forward Bulkhead Liner-Removal (P/B 401)
AMM 25-52-16-XXX-XXX	Forward Cargo Compartment Forward Bulkhead Liner-Installation (P/B 401)
AMM 52-31-00-XXX-XXX	Open the Cargo Door (P/B 201)
AMM 52-31-00-XXX-XXX	Open the Cargo Door (P/B 201)

TASK 21-27-01-XXX-XXX

1. Equipment Cooling Air Filter Removal

(Figure 1)

A. General

(1) You must remove electrical power from the airplane before you remove an equipment cooling air filter. This will make sure that electrical/electronic equipment does not receive electrical power when the equipment cooling system is not in operation.

B. Preparation for the Removal

SUBTASK 21-27-01-XXX-XXX

(1) Do this task: Remove Electrical Power, AMM TASK 24-22-00-XXX-XXX.

MAKE SURE YOU REMOVE ELECTRICAL POWER FROM THE AIRPLANE. IF YOU SUPPLY ELECTRICAL POWER TO THE ELECTRICAL/ELECTRONIC EQUIPMENT WHEN THE EQUIPMENT COOLING SYSTEM IS NOT IN THE OPERATION, THE ELECTRICAL/ ELECTRONIC EQUIPMENT CAN BECOME TOO HOT. THIS CAN CAUSE DAMAGE TO THE ELECTRICAL/ ELECTRONIC EQUIPMENT.

SUBTASK 21-27-01-XXX-XXX

(2) Open the forward cargo compartment door (Open the Cargo Door, AMM TASK 52-31-00-XXX-XXX).

SUBTASK 21-27-01-XXX-XXX

(3) Remove the forward right bulkhead liner in the forward cargo compartment. To remove the liner, do this task: Forward Cargo Compartment Forward Bulkhead Liner-Removal, AMM TASK 25-52-16-XXX-XXX.

C. Equipment Cooling Air Filter Removal

SUBTASK 21-27-01-XXX-XXX

(1) Close the forward cargo compartment door (Close the Cargo Door, AMM TASK 52-31-00-XXX-XXX).

NOTE The cargo door must be lowered to permit the removal of the filter.

SUBTASK 21-27-01-XXX-XXX

(2) Release the latch assemblies [3] that are on the top and the bottom of the filter housing assembly [1].

SUBTASK 21-27-01-XXX-XXX

(3) Removed the equipment cooling air filter [2] from filter hosing assembly[1].

SUBTASK 21-27-01-XXX-XXX

(4) If necessary, open the forward cargo compartment door (Open the Cargo Door, AMM TASK 52-31-00-XXX-XXX).

Words & Phrase		
	cause ~ ~을 일으키다. ~하게 하다	**cooling** 냉각
	electrical/electronic equipment 전기/전자 장비	
	electrical power 전력	**equipment** 장비품
	not in operation 작동하지 않음	**preparation** 준비
	remove 제거하다	**make sure that~** 확실하게 ~하다

Translation

1. 장비 냉각 공기 필터 장탈

A. 일반사항

(1) 장비 냉각 공기 필터를 제거하기 전에 비행기에서 전력을 제거해야 합니다. 이렇게 하면 장비 냉각 시스템이 작동하지 않을 때는 전기/전자 장비가 전력을 공급받지 않게 한다.

B. 장탈 준비

(1) 항공기의 전력을 제거하라 (AMM TASK 24-22-00-XXX-XXX)

CAUTION 〈주의〉 비행기에서 전력을 제거했는지 확인하라. 전기/전자 기기 냉각 시스템이 작동하지 않을 때 전기/전자 기기에 전원을 공급하는 경우, 전기/전자 장비가 너무 뜨거워질 수 있다. 이로 인해 전기/전자 장비가 손상될 수 있다.

(2) 전방 화물 실의 도어를 열어라. (AMM TASK 52-31-00-XXX-XXX)

(3) 전방 화물 실에서 전방 우측의 벌크헤드 라이너를 장탈하라. 라이너를 장탈하려면 AMM TASK 25-52-16-XXX-XXX의 작업을 수행하라.

C. 장비 냉각 공기 필터 장탈

(1) 전방 화물 실의 도어를 닫아라. (AMM TASK 52-31-00-XXX-XXX)

참고 필터를 장탈하기 위해서는 화물실의 도어를 내려야 한다.

(2) 필터 하우징 에셈블리[1]의 상부와 하부에 있는 래치 어셈블리[3]를 풀어 주어라.

(3) 필터 하우징 어셈블리[1]에서 장비 냉각 공기 필터[2]를 장탈하라.

(4) 필요하면, 전방 화물 실의 도어를 열어라. (AMM TASK 52-31-00-XXX-XXX)

TASK 21-27-01-XXX-XXX

2. Equipment Cooling Air Filter Installation
   (Figure 1)
   A. Expendable/Parts

AMM Item	Description	AIPC Reference	AIPC Effectivity
2	Filter	21-27-51-XX-XXX	ABC 002, 003, 331, 534, 802, 812-999
		21-27-51-XX-XXX	ABC 004, 005, 172, 204, 222, 271, 499, 569, 587
		21-27-51-XX-XXX	ABC 001, 013, 015, 589,
		21-27-51-XX-XXX	ABC 172,

B. Preparation for the Installation

SUBTASK 21-27-01-XXX-XXX

(1) Open the forward cargo compartment door (Open the Cargo Door, AMM TASK 52-31-00-XXX-XXX).

C. Equipment Cooling Air Filter Installation

SUBTASK 21-27-01-XXX-XXX

(1) Get access to the filter housing assembly [1] at the right, forward bulkhead in the cargo compartment.

SUBTASK 21-27-01-XXX-XXX

(2) Close the forward cargo compartment door (Close the Cargo Door, AMM TASK 52-31-00-XXX-XXX) to allow access for the installation of the air filter.

SUBTASK 21-27-01-XXX-XXX

(3) Put the equipment cooling air filter [2] in its position on the filter housing assembly [1].

SUBTASK 21-27-01-XXX-XXX

(4) Close the latch assemblies [3] that are on the top and bottom of the filter housing assembly [1].

D. Put the Airplane Back to its Usual Condition.

SUBTASK 21-27-01-XXX-XXX

(1) Install the forward right bulkhead liner in the forward cargo compartment. To Install the liner, do this task: Forward Cargo Compartment Forward Bulkhead Liner-Installation, AMM TASK 25-52-16-XXX-XXX.

SUBTASK 21-27-01-XXX-XXX

(2) Open the forward cargo compartment door to allow technician to exit.

SUBTASK 21-27-01-XXX-XXX

(3) As required, close the forward cargo compartment door.

Words & Phrase

- **as required** 필요한 만큼
- **AIPC Reference** 항공기 부품 도해 목록(Aircraft Illustrated Parts Catalog) 근거
- **bulkhead** 칸막이 벽(벌크헤드)
- **effectivity** 유효성(적용)
- **get access to~** ~에 접근하다
- **preparation** 준비
- **description** 명칭
- **expendable** 소모품
- **reference** 근거
- **put airplane back to~** 항공기를 ~로 되돌리다

Translation

2. 장비 냉각 공기 필터 장착

A. 소모성 자재/부품

AMM 항목번호/명칭/AIPC 근거/AIPC 적용

B. 장착 준비

(1) 전방 화물 실의 도어를 열어라. (AMM TASK 52-31-00-XXX-XXX)

C. 장비 냉각 공기 필터 장착

(1) 화물 실에서 우측 전방 벌크헤드에 있는 필터 하우징 어셈블리[1]에 접근하라.

(2) 에어 필터의 장착을 위한 접근할 수 있도록 전방 화물 실 도어를 닫아라. (AMM TASK 52-31-00-XXX-XXX)

(3) 장비 냉각 공기 필터[2]를 필터 하우징 어셈블리[1]의 제 위치에 집어넣어라.

(4) 필터 하우징 어셈블리[1]의 상부와 하부에 있는 래치 어셈블리[3]를 잠가 주어라.

D. 항공기의 원상 복구

(1) 전방 화물 실에서 전방 우측의 벌크헤드 라이너를 장착하라. 라이너를 장착하려면 AMM TASK 25-52-16-XXX-XXX의 작업을 수행하라.

(2) 작업자가 나올 수 있게 전방 화물 실 도어를 열어라.

(3) 필요하면 전방 화물 실 도어를 닫아라.

4.3 Practice Quiz

Answer Keys p. 254

01 Maintenance Manual, Task Card에 사용하는 기술 영어의 특징이 아닌 것은?

a. 작업 절차에는 명령문을 사용한다.
b. 수동태의 문장을 주로 사용한다.
c. 동사는 현재형, 과거형, 미래형만 사용한다.
d. 문장은 단어 수를 제한하여 짧게 구성한다.

02 기술 영어 문장에 대해 바르게 설명하지 않은 것은?

a. 작업 대상의 조건(형태, 상태)을 설명하는 문장은 "If~"로 시작한다.
b. 특정 전치사(to, for, with 등)만 사용한다.
c. "that ~"로 시작하는 문장은 대부분 작업의 내용을 설명하는 것이다.
d. "~ which ~"로 시작하는 문장은 대부분 작업 대상을 설명하는 것이다.

03 다음의 Lubrication 작업 문장에서 작업을 하기 위한 준비 조건에 해당하는 부분은?

> "Lubricate the No.4 and No.5 flap ballscrew nut, universal joint and jimbal cap with flaps in the 30-unit detent position, using a non-pneumatic hand-held grease gun and BMS 3-33 grease."

a. Lubricate
b. the No.4 and No.5 flap ballscrew nut, universal joint and jimbal cap
c. flaps in the 30-unit detent position
d. a non-pneumatic hand-held grease gun and BMS 3-33 grease

04 다음은 Detailed Inspection 내용이다. 결함을 발견할 수 있는 곳이 아닌 것은?

> "If bright metal particles or larger, non-metallic, black colored, fiber-like pieces are present in considerable quantity / on the filter element, in the filter cavity, or in the oil drained from the IDG, / do these steps."

a. considerable quantity
b. filter element
c. filter cavity
d. oil drained

05 다음은 Operational Check 내용이다. 내용을 바르게 설명하지 않은 것은?

> "Supply electrical power (AMM 24-22-00/201), Check the EICAS Status message ALTN VENT RELAY / does not show / on the lower EICAS display."

a. 전력을 공급하는 것은 EICAS를 작동하기 위함이다.
b. EICAS Status message가 보이지 않아야 정상이다.
c. "ALTN VENT RELAY" message가 보이지 않아야 정상이다.
d. 아래쪽에 있는 EICAS display에서 정상 작동 여부를 확인할 수 있다.

06 다음은 Functional Check 내용이다. 내용을 바르게 설명하지 않은 것은?

> "Check the tire pressure value shown for each tire on the TBMS (Tire Pressure and Brake Monitoring System) is within +/-10 psig of the tire pressure value measured with the tire pressure gauge at the tire."

a. 점검 목적은 TBMS의 기능 점검에 있다.
b. 본 기능 점검에서 기준이 되는 것은 Tire Pressure Gage에서 측정한 값이다.
c. TBMS 지시 값이 실제 측정한 값의 +/-10 psig 범위 안에 있으면 기능은 정상이다.
d. Tire Pressure Gage는 Tire에 장착되어 있다.

07 Task Card의 양식의 항목을 바르게 설명하지 않은 것은?

a. TITLE: Task Card의 제목
b. TASK: Task의 종류
c. WORK AREA: Task의 작업 부위
d. ACCESS: Task에 관련된 점검창의 개수

08 Task Card에서 "References" 소개된 AMM 번호와 제목은 무엇인가?

a. Task 작업할 때 참고해야 하는 내용
b. Task 작업을 수행하려면 발생하는 관련 작업
c. Task 작업을 할 때의 주의 사항
d. Task 작업을 위한 준비 사항

09 Task Card의 본문에서 처음에 소개되는 "General"은 어떤 내용을 담고 있나?

a. 본 작업을 할 때 준수하거나 참고할 사항
b. 본 작업의 주요 내용이나 사항
c. 본 작업의 수행에 필요한 시간
d. 본 작업의 수행 시 지켜야 하는 경고 사항

10 Task Card에서 사용하는 경고, 주의, 참고 사항 중 해당 절차의 바로 뒤에 소개되는 것은?

a. Warning
b. Caution
c. Note
d. Remark

11 Task Card의 절차에서 "Put the Airplane Back to its Usual Condition"의 의미는?

a. 항공기에 연결되었던 외부 전원을 차단
b. 항공기의 조종 계통을 전부 중립 위치로 변경
c. 항공기를 후진시켜서 원래의 위치로 이동
d. 작업을 위해 변경시켰던 항공기의 상태를 작업 전의 상태로 복구

PART 4.4 Maintenance Records

4.4.1 기술 영어의 작성을 위한 국제 표준(ASD-STE100) 소개

ASD-STE100은 항공 산업계에서 영어로 기술 도서를 작성할 때, 사용하는 단어의 범위를 제한하고 문장의 구조도 간결하게 구성하도록 함으로써 비영어권의 업무 종사자도 기술 도서를 쉽고 정확하게 이해하도록 하기 위한 목적으로 제정한 표준으로 기술 영어 작성 지침과 사용을 권고하는 단어들을 정리한 사전으로 구성되어 있다.

이 표준은 항공기 기술 도서(정비 매뉴얼 등)에 사용된 기술 영어의 이해에 큰 도움이 되는 자료로 여기에서는 주요 내용만 간략하게 소개하고자 한다.

[그림 4-6]
Specification ASD-STE100의 표지

1) ASD-STE100 (Simplified Technical English)의 개발 배경

영어가 모든 산업 분야에서 공용 언어로 넓게 사용되고 있고, 특히 항공우주 및 방위 산업 분야의 기술 도서는 거의 모두 영어로 작성되고 있으나, 사용자의 언어 분포를 보면 영어권보다는 비영어권이 더 많이 차지하고 있다. 한정된 영어 지식이 있는 사용자가 기술 도서를 이해하는 데에 있어 커다란 장애 요소는 복잡한 문장의 구조와 의미가 비슷한 단어(동의어)들의 무분별한 사용이다.

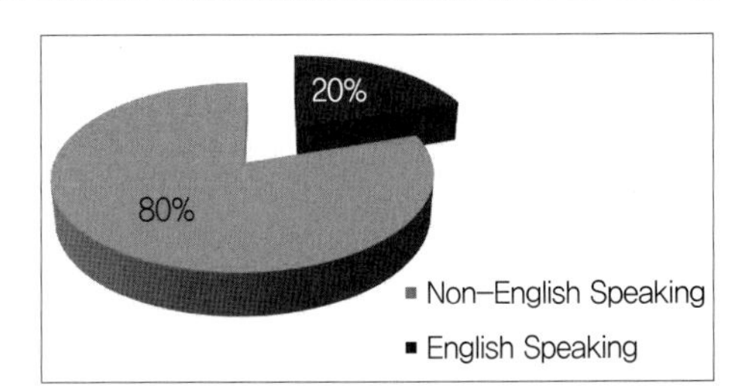

[그림 4-7] 항공우주 및
방위 산업 분야 종사자의 사용 언어 분포

더욱이 항공기의 정비를 하는 현장 작업자는 안전 운항을 위해서는 항공기의 모든 계통을 제작사의 기술 도서에 따라 정확하게 정비하여야 하는데, 이들이 해당 기술 도서의 내용을 완벽하게 이해하지 못한다면 이것은 매우 심각한 안전 저해 요소라 할 수 있다.

1970년대 후반부터 유럽항공사협회(AEA)는 민간 항공 산업에서 사용하는 영어 기술 도서를 현장 작업자가 사용하는 데 있어 무엇이 장애 요소인지 연구 과제로 선정하고 유럽 항공우주산업협회(AMCMA)로 하여금 보다 이해하기 쉬운 기술 도서가 발행될 수 있게 제도적인 장치를 마련하도록 요청하였다.

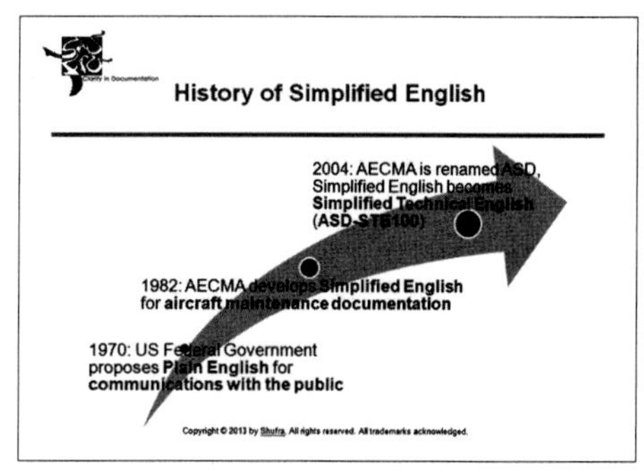

[그림 4-8] 간편 기술 영어(ASD-STE100)의 개발 역사

이러한 시대적 요구를 제안받은 그 당시 유럽항공우주산업협회(AMCMA:지금은 ASD)는 미국의 항공우주산업협회(AIA)와 공동으로 항공기 정비 도서 및 문서에 사용하는 영어 문장과 단어 형태를 분석하였고, 항공 산업체의 모든 기술 분야에서 더 간편하게 사용할 수 있도록 영어 문장의 구조와 형식을 단순화하고 사용하는 단어도 제한하는 작업을 하였다.
그 결과물을 "간편 기술 영어(STE: Simplified Technical English)"라고 호칭하고 모든 관련 산업체가 기술 도서를 작성할 때 적용하도록 "ASD-STE100"을 제정하였다.

2) 간편 기술 영어(ASD-STE100)의 구성과 주요 내용

Part 1: 작성 규칙(Writing Rules)

- Section 1 단어: 사전에 소개된 단어를 용도에 맞게 사용할 것.
- Section 2 명사: 명사는 세 단어 이하로 구성할 것.
- Section 3 동사: 문장의 동사는 단순형(현재, 과거, 미래형)만 사용할 것.
- Section 4 문장: 문장은 가능한 능동태로 사용하고 한 구절은 6개 문장 이하로 구성하며 기술 용어가 아니면 현재 완료형, 동명사형을 사용하지 말 것.
- Section 5 절차 작성: 절차는 단계별로 개별 문장으로 쪼개어 구성할 것.
- Section 6 설명 작성: 지침은 가능한 구체적으로 작성.
- Section 7 안전 지침: 안전 지시(경고 또는 주의)를 제일 먼저 기술하되 명확하고 단순한 명령문을 사용할 것.
- Section 8 단어 구성: 절차/설명 문장은 20/25개 단어 이하로 구성할 것.
- Section 9 작성 실습

Part 2: 단어사전(Dictionary)

- 현재 약 940여 개의 단어가 뜻과 용도(명사, 동사 등)와 함께 수록되어 있다.

3) 간편 기술 영어(ASD-STE100)의 공식 사용 기관 및 단체

- 유럽항공우주산업협회(ASD) 회원사
  (ASD:AeroSpace and Defence Industries Association of Europe)
- 미국 및 캐나다 항공우주산업협회(AIA) 회원사
  (Aerospace Industries Association of America and Canada)
- 항공우주산업협회의 국제협력위원회(ICCAIA) 회원사
  (International Coordinating Council of Aerospace Industries Associations
- 위에 소개한 협의의 고객인 항공사(Airlines)
- ASD 회원국과 미국의 방위부
  (Ministries of Defense, Department of Defense)
- A4A(Airlines for America 미국항공운송협회) 항공사
- 감항 당국(Aviation Authority)

4) 간편 기술 영어(ASD-STE100)에 대한 질의/응답

Q. 간편 기술 영어(STE)를 구술 언어로도 표준화하여 사용하여야 하나?
A. 아니다. 이 단어사전은 기술 문서를 작성하는 용도로만 사용하도록 개발된 것이다. 그러나 간편 기술 영어의 법칙과 제한된 단어를 회의나 세미나 같은 발표회에서 적용한다면 보다 많은 사람의 의사소통에 도움이 될 것이다.

Q. 간편 기술 영어(STE)는 문서 형태(text)도 규정하는가?
A. 아니다. 문장을 표현하는 방법만 규정하고 어떤 형태로 구성해야 한다는 규칙은 없다.

Q. 간편 기술 영어(STE)는 치수와 단위도 규정하는가?
A. 아니다. 기술 도서나 국가에 따라 단위나 치수는 자유롭게 선택하여 사용할 수 있다.

Q. 간편 기술 영어(STE)는 영어 문장 작성에 교육용으로 사용할 수 있는가?
A. 아니다. 간편 기술 영어는 언어교육용으로 개발된 것이 아니고 문장을 작성할 때 가급적 단순화하여 이해하기 쉬운 구조를 사용토록 규정한 것이다. 따라서 문장을 보다 명료하게 작성하기 위해서는 상대적으로 언어 수준이 높아야 하니 간편 기술 영어의 품질 향상을 위한다면 별도의 영어 능력을 키워야 한다.

Q. 간편 기술 영어(STE)는 기술 문서 번역에 도움이 되나?

A. 그렇다. 간편 기술 영어를 제정할 때 주요 목적이 번역을 쉽게 하기 위한 것이다. 따라서 용어 단어, 단어의 뜻, 문장 형태 등을 제한함으로써 문장의 변화를 최소화했기 때문에 자신의 모국어로 번역할 때 상대적으로 쉬울 뿐만 아니라 기계적인 번역기를 사용할 때에도 크게 도움이 된다.

Q. 간편 기술 영어(STE)에 대한 특별한 교육 방안은 있나?

A. 이 규정을 만든 항공우주방위산업협회(ASD)는 기본적으로 교육에 대한 별도의 방안을 가지고 있지 않다. 누구나 자유스럽게 교육 방안을 마련할 수 있다.

Q. 간편 기술 영어(STE)는 변화가 없는가?

A. 간편 기술 영어도 살아있는 언어이기 때문에 계속해서 변화가 이루어져야 한다. 기본적으로 개정은 3~4년 간격으로 이루어지며 개정된 내용은 STEMG web site (stemg@asd-ste100.org)를 통해서 최신판을 확인할 수 있다. 또한 사용하면서 문제점이 노출되거나 변경이 요구된다면 언제라도 자유스럽게 ASD STEMG (info@asd-ste100.org)에 제안할 수 있다. 참고로 현재의 규정은 2021년도에 발행된 제8판이다.

5) 간편 기술 영어(ASD-STE100) 자료의 획득 방법

간편 기술 영어(STE)의 모든 내용은 STEMG web site를 통하여 최신판 파일을 무료로 다운로드 받을 수 있다.

4.4.2 List of The Most "Popular" Recurring Errors

The table that follows gives you a list of the most "popular" recurring errors that writers do when they use STE.

If a word is not approved in the dictionary, do not use it.

NOTE

adj adjective 형용사
n noun 명사
TN technical name
adv adverb 부사
pron pronoun 대명사
v verb 동사
conj conjunction 접속사
prep preposition 전치사

Non-STE (잘못된 사용 사례)		STE (사전에 소개된 사용 권고 단어)
acceptable (adj)	다른 형용사로 대체	PERMITTED (adj)
alternate (adj)	"대체"의 의미가 아님	ALTERNATIVE (adj)
any (adj)	형용사로 사용 불필요	None or a different construction
avoid (v)	다른 동사로 대체	PREVENT (v)
both (adj)	"양쪽의" 형용사로 사용 금지	THE TWO (TN)
check (v)	동사로 사용 금지	CHECK (n)
cover (v)	동사로 사용 금지	COVER (TN)
damage (v)	동사로 사용 금지	DAMAGE (n)
ensure (v)	다른 동사로 대체	MAKE SURE (v)
fit (v)	다른 동사로 대체	INSTALL (v)
follow (v)	다른 동사로 대체	OBEY (v)
further (adj)	다른 형용사로 대체	MORE (adj)
further (adv)	다른 부사로 대체	MORE (adv)
have to (v)	다른 조동사로 대체	MUST (v)
insert (v)	다른 동사로 대체	PUT (v)
main (adj)	다른 형용사로 대체	PRIMARY (adj)
may (v)	다른 동사로 대체	CAN (v)
need (v)	동사로 사용 금지	NECESSARY (adj)

now (adv)	다른 문장으로 대체	AT THIS TIME
over (prep)	다른 전치사로 대체	ABOVE (prep), ON (prep), ALONG (prep)
perform (v)	다른 동사로 대체	DO (v)
press (v)	다른 동사로 대체	PUSH (v)
reach (v)	다른 동사로 대체	GET (v)
repeat (v)	다른 동사로 대체	DO (v) … AGAIN
required (v)	다른 형용사로 대체	NECESSARY (adj)
rotate (v)	다른 동사로 대체	TURN (v)
secure (v)	다른 동사로 대체	ATTACH (v), SAFETY (v)
shall (v)	다른 조동사로 대체	MUST (v)
should (v)	다른 조동사로 대체	MUST (v)
since (conj)	다른 접속사로 대체	BECAUSE (conj)
test (v)	동사로 사용 금지	TEST (n)
therefore (adv)	다른 부사로 대체	THUS (adv), AS A RESULT

4.4.3 Typical Sentences without and with STE applied

Non-STE A value of 2 mm is acceptable.
STE A VALUE OF 2 mm IS PERMITTED.
2mm 값이 허용된다.

Non-STE An alternate repair is available.
STE AN ALTERNATIVE REPAIR IS AVAILABLE.
대체 수리가 가능하다.

Non-STE Remove any of the four bolts.
STE REMOVE ONE OF THE FOUR BOLTS.
볼트 4개 중 하나를 장탈하라.

Non-STE Turn the controls slowly to avoid damage.
STE TURN THE CONTROLS SLOWLY TO PREVENT DAMAGE.
손상을 방지하려면 컨트롤을 천천히 돌려라.

Non-STE Attach both ends of the hose.
STE ATTACH THE TWO ENDS OF THE HOSE.
호스의 양쪽 끝을 연결한다.

Non-STE Check the valve for leakage.
STE DO A LEAKAGE CHECK OF THE VALVE.
밸브의 누설 여부를 점검하라.

Non-STE Cover the container.
STE PUT THE COVER ON THE CONTAINER.
용기에 덮개를 씌워라.

Non-STE Disconnect the spring so as not to damage the rod.
STE DISCONNECT THE SPRING TO PREVENT DAMAGE TO THE ROD.
로드가 손상되지 않도록 스프링을 분리하라.

Non-STE Ensure that the correct seals are installed
STE MAKE SURE THAT THE CORRECT SEALS ARE INSTALLED.
올바른 씰이 장착되었는지 확인하라.

Non-STE Fit the duct.
STE INSTALL THE DUCT.
덕트를 장착하라.

Non-STE When you use this material, follow the manufacturer's instructions.
STE WHEN YOU USE THIS MATERIAL, OBEY THE MANUFACTURER'S INSTRUCTIONS.
이 물질을 사용할 때는 제조업체의 지침을 따르라.

Non-STE If required, add further gaskets.
STE IF NECESSARY, ADD MORE GASKETS.
필요한 경우 개스킷을 추가하라.

Non-STE Cable tension can be further reduced if necessary by means of the turnbuckle.
STE IF IT IS NECESSARY TO DECREASE THE CABLE TENSION MORE, USE THE TURNBUCKLE.
케이블 장력을 더욱 줄일 필요가 있는 경우에는 턴버클을 사용하라.

Non-STE You have to use ear protection when you are near an engine that is in operation.
STE YOU MUST USE EAR PROTECTION WHEN YOU ARE NEAR AN ENGINE THAT IS IN OPERATION.
작동 중인 엔진 근처에 있을 때는 귀마개를 반드시 착용하라.

Non-STE Insert the sleeve into the opening.
STE PUT THE SLEEVE INTO THE OPENING.
슬리브를 개구부에 삽입하라.

Non-STE The main cause of valve failure is contamination of hydraulic fluid.
STE THE PRIMARY CAUSE OF VALVE FAILURE IS CONTAMINATION OF THE HYDRAULIC FLUID.
밸브 고장의 주요 원인은 유압유의 오염이다.

Non-STE The vanes may be damaged by using incorrect equipment.
STE IF YOU USE INCORRECT EQUIPMENT, YOU CAN CAUSE DAMAGE TO THE VANES.
잘못된 장비를 사용하면 베인이 손상될 수 있다.

Non-STE The backing rings do not need to be replaced.
STE IT IS NOT NECESSARY TO REPLACE THE BACKING RINGS.
백킹 링은 교체할 필요가 없다.

Non-STE Do not tighten the nuts now.
STE DO NOT TIGHTEN THE NUTS AT THIS TIME.
지금은 너트를 조이지 마라.

Non-STE Make sure the hydraulic fluid level is not over the FULL mark.
STE MAKE SURE THAT THE HYDRAULIC FLUID LEVEL IS NOT ABOVE THE "FULL" MARK.
유압유의 레벨이 "FULL" 표시를 넘지 않았는지 확인하라.

Non-STE Install the stop sleeve over the sliding member
STE INSTALL THE STOP SLEEVE ON THE SLIDING MEMBER.
슬라이딩 부재 위에 스톱 슬리브를 장착하라.

Non-STE The weight must be evenly spread over the stabilizer span.
STE APPLY THE WEIGHT EQUALLY ALONG THE STABILIZER SPAN.
스태빌라이저의 스팬을 따라서 균일하게 무게를 얹어 주어라.

Non-STE Perform the leak test.
STE DO THE LEAK TEST.
누설 테스트를 수행하라.

Non-STE Press and hold the TEST button.
STE PUSH AND HOLD THE TEST BUTTON.
테스트 버튼을 누르고 있어라.

Non-STE When the correct pressure is reached, close the valve.
STE WHEN YOU GET THE CORRECT PRESSURE, CLOSE THE VALVE.
올바른 압력에 도달하면 밸브를 닫아라.

Non-STE Repeat steps (10) to (14).
STE DO STEPS (10) THRU (14) AGAIN.
(10) ~ (14) 단계를 반복해서 수행하라.

Non-STE Install clean filters if required.
STE IF NECESSARY, INSTALL CLEAN FILTERS.
필요한 경우 깨끗한 필터를 장착하라.

Non-STE Slowly rotate the vane.
STE SLOWLY TURN THE VANE.
베인을 천천히 회전시켜라.

Non-STE Holes shall not have sharp edges.
STE HOLES MUST NOT HAVE SHARP EDGES.
구멍에는 날카로운 모서리가 있어서는 안 된다.

Non-STE Personnel should wear protective clothing.
STE PERSONNEL MUST USE PROTECTIVE CLOTHING.
작업자는 보호복을 착용해야 한다.

Non-STE Since Alodine is a dangerous material, be careful when you use it.
STE BE CAREFUL WHEN YOU USE ALODINE, BECAUSE IT IS A DANGEROUS MATERIAL.
알로다인은 위험물질이므로, 사용 시 주의하라.

Non-STE Functionally test warning system.
STE DO A FUNCTIONAL TEST OF THE WARNING SYSTEM.
경고 시스템의 기능 시험을 하라.

[Words and Phrases]

acceptable 허용되는, 받아들일 수 있는
available 이용할 수 있는, 사용 가능한
alternate 대체의, 교체의, 번갈아 하는
avoid 피하다, 회피하다
by means of ~에 의하여
correct 수정하다, 교정하다
disconnect 분리하다
if required 필요하다면
leak test 누설 시험
protection 보호
tighten 죄다, 팽팽하게 치다
cause 원인, 이유
damage 손상, 피해
ensure 보증하다, 책임지다
incorrect 부정확한, 틀린
leakage 누설, 누수
spread over 퍼지다
contamination 오염, 더러움
dangerous 위험한
further 다시, 더 나아가
instructions 지시, 명령, 설명서
perform 수행하다
tension 장력

4.4.4 Writing Rules of Maintenance Records

Maintenance Records are evidence of maintenance work, therefore the contents should be concise but clear and specific. It is recommended to apply the rules of Simplified Technical English in writing sentences as much as possible.

Words & Phrase

clear 명확한	**concise** 간결한	**contents** 내용
evidence 증거	**as much as possible** 최대한, 가능한 한 많이	
is (be) necessary to~ ~하는 것이 필요하다.		**rules** 규칙
specific 구체적인	**therefore** 그러므로	

Translation

정비 기록은 정비 작업의 증거 자료이므로 내용이 간결하되 명확하고 구체적이어야 한다. 문장을 작성할 때 Simplified Technical English의 규칙을 최대한 적용하는 것이 좋습니다.

참고 항공기 제작사는 STE의 작성 규칙을 100% 정비 도서에 반영하고 있지는 않으므로, 정비 기록을 작성할 때 일부 동사나 명사는 정비 도서에서 사용하는 것을 그대로 이용할 수도 있다.

1) Write in capital letters in English.
(영어로 대문자로 작성하라.)

사례 removed and stowed all L/G ground lock pins.
권고 REMOVED AND STOWED ALL L/G GROUND LOCK PINS.

2) Use past tense for verbs, but use simple words with the same meaning whenever possible.
(동사는 과거형을 사용하되 같은 의미의 단어 중에서 간단한 것을 사용하라.)

사례 REMOVED AND REPLACED (WITH) NEW SPRING
권고 REPLACED THE SPRING.

3) Use simple forms of prepositions whenever possible.
(전치사는 가능하면 간단한 것을 사용하라.)

사례 AS PER, IN ACCORDANCE WITH, I/A/W, IAW AMM 25-25-07
권고 PER AMM 25-25-07

4) Record the type and number of reference documents of maintenance work performed.
(수행한 정비 작업의 근거가 되는 문서의 종류와 번호를 기록하라.)

사례 REPLACED THE AIR CHILLER.
권고 REPLACED THE AIR CHILLER PER AMM 25-31-20.

5) Use only abbreviations and acronyms introduced in the maintenance manual, but do not create them arbitrarily.
(약어는 정비 도서에 소개된 것만 사용하되 임의로 만들어 사용하지 마라.)

사례 RPLCD THE FLTR ELEMENT.
권고 REPLACED THE FILTER ELEMENT.

6) Write the contents clearly and specifically (replace, repair, inspection, test, modification, etc.).
(내용은 구체적으로 명확하게 기록하라: 교환, 수리, 검사, 시험, 개조 등으로 구분)

사례 REPLACED THE COFFEE MAKER AND CHECK NORMAL.
권고 REPLACED THE COFFEE MAKER PER AMM 25-31-00 AND OPERATIONAL CHECK FOUND NORMAL.

7) Do not use nouns as verbs or verbs as nouns.
(명사를 동사로 사용하거나 동사를 명사로 사용하지 마라.)

사례 REPLACED THE AIR CHILLER AS PER AMM 25-33-02 AND OPER CHECKED NORMAL.
권고 REPLACED THE AIR CHILLER PER AMM 25-33-02 AND OPERATIONAL CHECK FOUND NORMAL.

8) Use the same words such as verbs that are used in the manual.
(동사 등의 단어는 Manual에서 사용하는 것과 같은 단어를 사용하라.)

사례 PULL OUT (PUSH IN) THE CIRCUIT BREAKER.
권고 OPEN (CLOSE) THE CIRCUIT BREAKER.

Manual 사용 단어 사례 ACCOMPLISHED THE TASK~,
PERFORMED THE LEAK TEST~ 등.

9) If possible, do not use more than 4 nouns in combination. However use the noun as is if that is introduced in the Manual.
(명사는 4개 이상을 붙여 사용하지 마라. 단 Manual에 소개된 명사는 그대로 사용하라.)

사례 THE FORWARD TURBINE OVERHEAT THERMOCOUPLE TERMINAL TAGS.

권고 THE TERMINAL TAGS ON THE FORWARD OVERHEAT THERMOCOUPLE OF THE TURBINE.

Manual 사용 명사 사례 MAIN GEAR DOOR RETRACTION WINCH HANDLE.

10) "Test" and "Check" are not used in combination.
(Test와 Check는 중복하여 사용하지 마라.)

사례 REMOVED AND REPLACED WITH NEW LAV DRAIN VALVE ASSY AND LEAK TEST CHECKED NORMAL AS PER AMM 38-32-63.

권고 REPLACED THE LAV DRAIN VALVE ASSY PER AMM 38-32-63 AND LEAK TEST FOUND NO LEAK.

11) Record the check result or test result in more detail than "NORMAL" if possible.
(Check 나 Test 결과는 가능하면 "NORMAL"보다 구체적으로 기록하라.)

사례 PERFORMED A NDT OF DENT AREA PER NTM PART 6 / 51-00-00 / PROCEDURE 23 AND FOUND NORMAL.

권고 PERFORMED A NDT OF DENT AREA PER NTM PART 6 / 51-00-00 / PROCEDURE 23 AND FOUND NO CRACK.

권고 NORMAL → CORRECT, NO LEAK, NO CRACK, NO DEFECT, NO CORROSION.

12) Long sentences must be written separately by content and divided into short sentences.

(긴 문장은 내용별로 구분하여 짧은 문장으로 나누어서 작성하라.)

사례 PERFORMED FIM 27-89-00. DM14700A WIRING CHECK BETWEEN PIN 3 AND PIN 6. 152.8Ω LIMIT OVER AND REMOVED AND REPLACED PRIMARY CIRCUIT TERMINAL 11 AND 4, 12 AND 5. OPERATION CHECK NORMAL. NO STATUS MSG AS PER AMM 27-89-04, WDM 27-89-11.

권고 1) PERFORMED FIM 27-89-00. A WIRING CHECK OF DM14700A BETWEEN PIN 3 AND 6 FOUND LIMIT OVER(152.8Ω).
2) REPLACED PRIMARY CIRCUIT TERMINAL 11 AND 4 / 12 AND 5.
3) PERFORMED A OPERATIONAL CHECK PER AMM 27-89-04 / WDM 27-89-11 AND FOUND NORMAL/NO STATUS MSG.

13) When connecting related sentences, use conjunctions such as "AND", "BUT", and "AND THUS".
(관련 문장을 연결할 때는 "AND", "BUT", "AND THUS" 등의 접속사를 사용하라.)

사례 REPLACED THE DRAIN VALVE. C/B POPPED-OUT DURING VALVE OPER CHECK. INSTALLED THE REMOVED DRAIN VALVE.

권고 REPLACED THE DRAIN VALVE BUT C/B POPPED-OUT DURING VALVE OPERATIONAL CHECK THUS INSTALLED THE REMOVED DRAIN VALVE.

14) When recording the position of parts/equipment, record it using the method used in the manual.
(부품/장비품의 위치(POSITION) 기록은 MANUAL에 명시된 방식으로 기록하라.)

사례 REPLACED #1 ENG AND PERFORMED OPERATIONAL CHECK PER XXXXX.
사례 REPLACED NBR.1 ENG AND PERFORMED OPERATIONAL CHK PER XXXXX.
권고 REPLACED No.1 ENG AND PERFORMED OPERATIONAL CHECK PER XXX.

15) Left, Right, Front, Rear, are recorded in one unified way.
(좌측, 우측, 전방, 후방 등은 한 가지로 통일하여 기록하라.)

사례 REMOVED LEFT(RIGHT) AILERON
사례 REMOVED L/H(R/H) AILERON
권고 REMOVED LH(RH) AILERON.

4.4 Practice Quiz

Answer Keys p. 254

01 ASD-STE100의 제정 배경을 바르게 설명한 것은?

a. 항공기 기술 도서는 제작사에 따라 다양한 언어로 작성된다.
b. 항공기 정비 분야의 종사자는 비영어권보다 영어권이 더 많다.
c. 항공기 기술 도서의 이해도는 항공기 안전 운항과 관련이 없다.
d. 항공기 기술 도서를 이해하는 데 주요 장애 요소는 복잡한 문장구조와 다양한 단어의 사용이다.

02 ASD-STE100의 내용을 바르게 설명한 것은?

a. 기술 도서의 내용을 이해하는 데 도움이 되는 내용을 담고 있다.
b. 기술 도서에서 사용하는 치수와 단위를 규정하는 내용이 있다.
c. 기술 도서의 형태를 규정하는 내용이 있다.
d. 기술 영어의 기본적인 작성 지침이므로 내용은 개정이 되지 않는다.

03 STE의 작성규칙(Writing Rules)을 바르게 설명하지 않은 것은?

a. 명사는 세 단어 이하로 구성한다.
b. 동사는 현재, 과거, 미래형만 사용한다.
c. 문장은 가급적 수동태 문장을 사용한다.
d. 문장은 절차/설명의 경우 20/25개 이하의 단어로 구성한다.

04 다음은 기술 영어 작성 시 사용한 단어를 STE에서 권고하는 단어로 교체한 사례이다. 바르게 소개하지 않은 것은?

a. ensure(동사) → make sure(동사)
b. avoid(동사) → prevent(동사)
c. fit(동사) → install(동사)
d. should(조동사) → shall(조동사)

05 다음 중 정비 기록의 중 작성 원칙이 아닌 것은?

a. 영문으로 대문자로 기록한다.
b. 정비 도서에는 소개되지 않았으나 일반적으로 통용되는 약어는 사용해도 좋다.
c. 작업 내용을 설명하는 동사는 과거형을 사용한다.
d. 내용은 간결하되 구체적이고 명확해야 한다.

06 다음의 정비 기록 중에서 작성 원칙에 맞게 작성된 것은?

a. REPLACED THE NOSE RADOME IN ACCORDANCE WITH AMM 53-XX-XX.
b. Reformed Borescope Inspection of No.3 ENG Combustion Chamber.
c. CLEAN LH OUTFLOW VALVE PER AMM 21-XX-XX.
d. REPAIRED THE L/E OF LH AILERON PER SRM 57-XX-XX.

07 다음의 정비 기록을 바르게 수정한 것은?

"PERFORM A NDT OF THE DENT AREA AS PER 747 NTM PART 6/51-00-00/PROCEDURE 23 AND FOUND NORMAL."

a. PERFORM A NDT OF THE DENT AREA PER 747 NM PART 6/51-00-00/ PROCEDURE 23 AND FOUND NO CRACK.
b. PERFORMED A NDT OF THE DENT AREA AS PER 747 NTM PART 6/51-00-00/ PROCEDURE 23 AND FOUND NORMAL.
c. PERFORM A NDT OF THE DENT AREA PER 747 NTM PART 6/51-00-00/ PROCEDURE 23 AND FOUND NO CRACK.
d. PERFORMED A NDT OF THE DENT AREA PER 747 NTM PART 6/51-00-00/ PROCEDURE 23 AND FOUND NO CRACK.

Answer Keys

PART 4.1 Aircraft Maintenance

01 b
02 c
03 b
04 d
05 b
06 c

PART 4.2 Maintenance Program

01 c
02 b
03 c
04 d
05 b
06 d
07 b
08 c

PART 4.3 Maintenance Task Card

01 b
02 b
03 c
04 a
05 b
06 d
07 d
08 b
09 a
10 c
11 d

PART 4.4 Maintenance Records

01 d
02 a
03 c
04 d
05 b
06 d
07 d

Appendix

부록

1 General Definitions
2 Verbs and Prepositions frequently used in Maintenance Manual
3 Abbreviations and Acronyms used in Aircraft Maintenance

Appendix 1

General Definitions

Aeroplane. A power-driven heavier-than-air aircraft, deriving its lift in flight chiefly from aerodynamic reactions on surfaces which remain fixed under given conditions of flight.

Air carrier. A person who undertakes directly by lease, or other arrangement, to engage in air transportation.

Aircraft. Any machine that can derive support in the atmosphere from the reactions of the air other than the reactions of the air against the earth's surface.

Airmanship. The consistent use of good judgement and well-developed knowledge, skills and attitudes to accomplish flight objectives.

Air operator certificate (AOC). A certificate authorizing an operator to carry out specified commercial air transport operations.

Airport. An area of land or water that is used or intended to be used for the landing and takeoff of aircraft, and includes its buildings and facilities, if any.

Air traffic. The aircraft operating in the air or on an airport surface, exclusive of loading ramps and parking areas.

Air traffic clearance. An authorization by air traffic control, for the purpose of preventing collision between known aircraft, for an aircraft to proceed under specified traffic conditions within controlled airspace.

Airworthy. The status of an aircraft, engine, propeller or part when if conforms to its approved design and is in a condition for safe operation.

Balloon. A lighter-than-air aircraft that is not engine driven, and that sustains flight through the use of either gas buoyancy or an airborne heater.

Cabin crew member. A crew member who performs, in the interest of safety of passengers, duties assigned by the operator or the pilot-in-command of the aircraft, but who shall not act as a flight crew member.

Certify as airworthy. To certify that an aircraft or parts thereof comply with current airworthiness requirements after maintenance has been performed on the aircraft or parts thereof.

Commercial air transport operation. An aircraft operation involving the transport of passengers, cargo or mail for remuneration or hire.

Competency. A dimension of human performance that is used to reliably predict successful performance on the job. A competency is manifested and observed through behaviours that mobilize the relevant knowledge, skills and attitudes to carry out activities or tasks under specified conditions.

Continuing airworthiness. The set of processes by which an aircraft, engine, propeller or part complies with the applicable airworthiness requirements and remains in a condition for safe operation throughout its operating life.

Cruising level. A level maintained during a significant portion of a flight.

Dangerous goods. Ariticles or substances which are capable of posing a risk to health, safety, property or the environment and which are shown in the list of dangerous goods in the Technical Instructions or which are classifed according to those Instructions.

Duty. Any task that flight or cabin crew members are required by the operator to perform, including, for examples, flight duty, administrative work, training, positiong and standby when it is likely to induce fatigue.

Engine. A unit used or intended to be used for aircraft propulsion. It consists of at least those components and equipment necessary for functioning and control, but excludes the propeller/rotors(if applicable).

Error. An action or inaction by an operational person that leads to deviations from organizational or the operational person's intentions or expectations.

Extended diversion time operations (EDTO). Any operation by an aeroplane with two or more turbine engines where the diversion time to an en-route alternate aerodrome is greater than the threshold time established by the state of the operator.

Flight plan. Specified information provided to air traffic services units, relative to an intended flight or portion of a flight of an aircraft.

Flight time. The total time from the moment an aeroplane first moves for the purpose of taking off until the moment it finally comes to rest at the end of the flight.

Human factors principles. Principles which apply to aeronautical design, certification, training, opeations and maintenance and which seek safe interface between the human and other system components by proper consideration to human performance.

In-flight shutdown(IFSD). When an engine ceases to function (when the airplane is airborne) and is shutdown, whether self induced, flightcrew initiated or caused by an external influence. The FAA considers IFSD for all causes: for example, flameout, internal failure, flightcrew initiated shutdown, foreign object ingestion, icing, inability to obtain or control desired thrust or power, and cycling of the start control, however briefly, even if the engine operates normally for the remainder of the flight.

Load factor. The ratio of a specified load to the total weight of the aircraft. The specified load is expressed in terms of any of the following: aerodynamic forces, inertia forces, or ground or water reactions.

Maintenance. The performance of tasks required to ensure the continuing airworthiness of an aircraft, including any one or combination of overhaul, inspection, replacement, defect rectification, and the embodiment of a modification or repair.

Maintenance release. A document which contains a certification confirming that the maintenance work to which it relates has been completed in a satisfactory manner, either in accordance with the approved data and the procedures described in the maintenance organization's procedures manual or under an equivalent system.

Master minimum equipment list (MMEL). A list established for a particular aircraft type by the organization responsible for the type design with the approval of the State of Design containing items, one or more of which is permitted to be unserviceable at the commencement of a fligth. The MMEL may be associated with special operating conditions, limitations or procedures.

Modification. A change to the type design of an aircraft, engine or propeller.

Operation specifications. The authorizations, conditions and limitations associated with the air operator certificate and subject to the conditions in the operations manual.

Operator. The person, organization or enterprise engaged in or offering to engage in an aircraft operation.

Pilot-in-command. The pilot designated by the operator, or in the case of general aviation, the owner, as being in command and charged with the safe conduct of a flight.

Powerplant. The system consisting of all the engines, drive system components (if applicable), and propellers (if installed), their acceessories, ancillary parts, and fuel and oil systems installed on an aircraft but excluding the rotors for a helicopter.

Preventive maintenance. Simple or minor preservation operations and the replacement fo small standard parts not involving complex assembly operations.

Rating. An authorization entered on or associated with a licence and forming part thereof, stating special conditions, privileges or limitations pertaining to such licence.

Repair. The restoration of an aircraft, engine, propeller or associated part to an airworthy condition in accordance with the appropriate airworthiness requirements after it has been damaged or subjected to wear.

Safety management system (SMS). A systematic approach to managing safety, including the necessary organizational structures, accountability, responsibilities, policies and procedures.

Type certificate. A document issued by a Contracting state to define the design of an aircraft, engine or propeller type and to certify that this design meets the appropriate airworthiness requirements of that State.

Type design. The set of data and information necessary to define an aircraft, engine or propeller type for the purpose of airworthiness determination.

Appendix 2

Verbs and Preposition frequently used in Maintenance Manual

1. activate: 작동시키다 (활성화시키다)

   Activate the standby hydraulic pumps.

2. adjacent to ~: ~ 근처에, ~에 인접한

   Only one bolt is used in the hole adjacent to the nameplate.

3. adjust ~: ~을 조절하다

   Adjust the tabs of the fuel filter element to the fuel filter cover.

4. alternates to ~: ~을 대체하는 것(~의 대체품)

   The tools shown are alternates to each other within the same airplane series.

5. applicable to ~: ~에 해당된다

   The test is applicable to the left and the right ATC system.

6. apply to ~: ~에 적용된다

   These steps apply to all the metal support equipment within a 50 ft radius of an open fuel tank.

7. apply A to (on) B: B 에 A 를 바르다 (칠하다)

   Do not apply the grease or oil to the stainless steel (CRES) control cables.

8. as follows: 다음과 같이, 다음 처럼

   Examine the retract actuator assembly as follows:

9. as required: 필요한 만큼

   Remove sidewalls and floor panels as required.

10. (be) associated with ~: ~ 와 관련된

Pressurize the applicable hydraulic system with the hydraulic pump that is associated with the installed filter element:

11. attach A to B: A 를 B 에 부착하다

Attach the fixture, SPL-3899, to the relief valve.

12. become ~: ~(워) 지다, …이 되다

Electronic equipment can become too hot.

13. between A to B: A 와 B 사이에

Make sure that there is no continuity between pins A2 and A1.

14. cause A to B: B 에 A 를 일으키다 (B 에 A 를 발생시키다)

This can cause injury to persons and damage to equipment.

15. change A to B: A 를 B 로 바꾸다(변환 하다)

Figure 3 may be used to change binary data to the correct engineering units.

16. A change(s) to B: A 가 B 로 바뀌다(변하다)

Make sure that the FLOW INDICATOR changes to yellow each time that you breathe.

17. check ~: ~을 점검하다

Check the hinge seal [2] on the door frame for these abnormal conditions:

18. clean A with B: A 를 B 로(를 사용하여) 세척하라

Clean the mating interfaces of the filter housing and the filter bowl with a cotton wiper.

19. close ~: ~을 닫다, ~을 잠그다

Close these circuit breakers.

20. A comes on while B: B 를 하는 중에 A 가 나타나다(들어오다)

Make sure that the code 90 comes on while the test is in progress.

21. A comes on to show B: B 가 되었음을 보여주는 A 가 나타나다(들어오다)

Make sure that the code 97 comes on to show the test is complete.

22. complete ~: ~을 완료하다 (~을 마치다, ~을 끝내다)

You will need 3 persons to complete this task.

23. connect A to B: A 를 B 에 연결하다

Connect the electrical connector D12156 to the captain's cutout switch.

24. continue to ~: 계속해서 ~하다 (지속적으로 ~하다)

Continue to push the relief valve door to the full open position.

25. A contains B about C: A 는 C 에 대한 B 를 담고 있다 (~이 들어 있다)

The TR-220 Operating Manual also contains information about the tests.

26. do the deactivation of ~: ~이 작동되지 않게 하다

Do the deactivation of the leading edge and the thrust reverser (for ground maintenance).

27. (be) diminished: 줄어들다, 약화되다

The cooling capability could be diminished.

28. disassemble~: ~을 분해(해체)하다

Do not disassemble connectors when you do this task.

29. discard ~: ~을 버리다(~을 폐기하다)

Discard all personal protective equipment after you use it one time.

30. disconnect A from B: B 로부터 A를 분리하다

Disconnect the electrical connectors from the squibs.

31. due to ~: ~로 인해

Make sure that the shielding at the connector has not degraded due to obvious damage, failure or irregularity.

32. during ~: ~하는 동안

Move the balance panel hinges during lubrication.

33. either A or B: A 또는 B

You can use either the drain coupling or the static port adapter to pressurize the static system.

34. (be) engaged at ~: ~에 맞물리다

Make sure that the coupling rings are engaged at their detent.

35. examine ~: ~을 자세히 살펴보다, ~을 조사하다

Examine the fusible tips on the discharge tubes.

36. except ~: ~을 제외한(제외하고)

Applicable to all filter basket part numbers except 01940-001 and 01956-000.

37. for ~: ~에, ~을 위해서, ~에 관하여, ~에 대한

Install the two screws for the retainer on the filter housing.

38. get (gain) access to ~: ~에 접근하라

Get (gain) access to the wheel well of the main landing gear.

39. hold ~: ~을 잡아주다, ~을 붙잡아 주라

Hold the retainer in its position on the filter housing.

40. if ~: ~하다면, ~인 경우에는

Remove electrical power if it is not necessary.

41. if installed: 장착되어 있다면, 장착된 경우에는

Remove the antenna shield, if installed.

42. if necessary: 필요하면

Remove the L/G down lock pins if necessary.

43. illuminate: 불이 들어오다, 밝아지다

Verify that the STBY RUD ON light on the overhead panel P5-3 is not illuminated.

44. in accordance with ~: ~에 따라

Washers may be installed for fastener grip length in accordance with SRM Chapter 51.

45. include ~: ~을 포함하다

Elapsed time does not include work preparation and cure time of chemical.

46. indicate ~: ~을 나타낸다, ~을 지시한다

Make sure that the display indicates SELF TEST PASS.

47. indicated in~: ~에 명시된

With the exception of the wire bundles indicated in the table below, the Engine Harnesses are considered LRUs and should not be repaired.

48. (be) inhibited: 억제되다, 작동이 안되게 하다

The E/E cooling supply low flow sensor warning is inhibited for 5 minutes.

49. (be) in progress: 진행 중인

Notify the local ATC that the transponder testing is in progress.

50. install ~: ~을 장착하다

Install the cover plate onto the door structure with the bolts and the washers.

51. in the event of ~: ~이 발생한(생긴) 경우

In the event of a forward cargo fire, the cargo smoke detectors will detect smoke.

52. keep ~: ~한 상태로 하다 (~한 상태를 유지하다)

Keep persons and equipment away from the flight control surfaces.

53. less than ~: ~ 보다 적은 (4개 이하, 5개는 포함되지 않음)

The trailing edge flaps are at less than 5 unit or more than 25 units.

54. look for ~: ~을 찾다 (~을 살펴보다)

Look for unwanted particles on the bearing surface.

55. loosen ~: ~을 풀다(풀어주다)

Loosen and remove the bolt and the flat washer that hold the fuel filter cover.

56. Lubricate A with B: A 를 B 로 윤활하다, A 에 B 를 발라 주어라

Lubricate a new packing with oil or Vaseline.

57. make a note of ~: ~을 기록하다

Make a note of that condition for future reference.

58. make sure that ~: ~을 확인하다,

Make sure that the down lock pins are installed on the nose and main landing gear.

59. to make sure ~: ~을 확인하기 위해

Perform a detailed visual inspection to make sure that pulleys are free to rotate.

60. may be ~ ed: ~해도 된다, ~ 할 수 있다

Washers may be installed for fastener grip length in accordance with SRM Chapter 51.

61. more than ~: ~보다 많은 (than 다음에 오는 숫자나 수량은 포함되지 않음)

Do not put a load of more than 65 pounds on the elevators at one time.

62. mount A on B: A 를 B 에 설치하다

Mount the Directional Antenna on the Test Sets friction hinge.

63. need ~: ~이 필요하다

You will need a small hand held mirror to see the forward face of the cooler.

64. notice ~: ~을 알게 되다

If you notice any damage along the length of the wire bundle, make a note and do any repairs.

65. obey ~: ~을 준수하다 (~을 반드시 따르다)

Obey these instructions to prevent injuries to personnel.

66. (be) obstructed: ~이 가로 막히다, ~의 진행이 방해를 받다

This indicates that airflow is significantly obstructed and the cooling capability could be diminished.

67. obtain ~: ~을 얻다, ~을 획득하다

If necessary, jack individual points to obtain the desired attitude.

68. open ~: ~을 열다

Open the forward cargo compartment door.

69. operate: ~을 작동하다

Operate the elevator in the full travel 6 times.

70. (be) operated: ~이 작동되다

The correct pump must be operated to do the leakage check.

71. operationally check ~: ~을 작동 점검하다

Operationally check the landing gear transfer valve.

72. override ~: ~을 건너뛰다 (~을 무시하다)

You can override and skip to the next test in the sequence by toggling the AUTO/TEST/MANUAL switch to the AUTO position.

73. perform ~: ~을 수행하다

Perform a general visual inspection of the eductor (on the APU) for general condition.

74. (be) permitted: 허용되다

Cracks less than 67% (2/3) of the bracket are permitted.

75. place A on B: A 를 B 에 놓다(비치하다)

Place a DO-NOT-MOVE tag on the control column.

76. (be) preceded by ~: ~이 앞서다(~가 앞에 나온다)

Tool part numbers that are replaced or non-procurable are preceded by "Opt:", which stands for Optional.

77. preparation for ~: ~의 준비, ~을 위한 준비

Preparation for Recirculation Fan Check Valve Inspection.

78. pressurize ~: ~을 가압하다, ~에 압력을 가하다

Pressurize the rudder hydraulic systems A and B.

79. prevent ~: ~을 방지하다

To prevent NLG movement, install a lockout pin in the nose wheel steering mechanism.

80. prior to ~: ~을 하기 전에, ~전에

All three detectors are examined for correct installation [locked] by two different persons prior to the close of the fan cowls.

81. pull ~: ~을 당기다

Pull the T-handle to close the relief valve manually.

82. push ~: ~을 누르다

Push the horn cutout switch.

83. put A to (in, into) B ~: A 를 B 의 상태로 만들다(만들어 주다)

Put the airplane in (into) the takeoff configuration.

84. put A back to B : A 를 B 의 상태로 되돌려 놓다

Put the airplane back to its usual condition.

85. put on ~: ~을 착용하다

Put on goggles, and gloves when you use fuel.

86. referred to (as) ~: ~로 부른다, ~라고 한다

For this task, the IDG air/oil cooler will be referred to as the cooler.

87. release ~: ~을 풀어 주다, ~을 풀어 내다

Turn the turnbuckles to release the tension of the control cables RA and RB.

88. remove A from B: B 로부터 A 를 장탈(제거)하다

Remove the four bolts of the test fixture from the access door.

89. repair ~: ~을 고치다(수리하다)

If necessary, do these steps to repair or replace the leveling compound.

90. (be) repaired: ~이 수리되다

Cracks cannot be repaired with the APU on the airplane.

91. repeat ~: ~을 반복 (수행) 하다

Repeat procedure a) through e) to opposite side of the airplane.

92. replace ~: ~을 교환하다

Repair or replace the check valves that do not meet the minimum thickness.

93. return A to B: A 를 B 로 되돌리다

Return the Airplane to the Ground Mode.

94. restore A to B: B 에 A 를 복구하다(회복시키다)

To restore pressure to the potable water system, do this task: Potable Water System – Activation.

95. rinse ~: ~을 씻다, ~을 헹구어 내다

If water is used, rinse the cooler using the pressure washer.

96. select ~: ~를 선택하다

Select the RTE function key on the FMC MCDU.

97. service ~: 보급하다, 보충하다

Check the oil level of the alternate flap drive gearbox and service as required.

98. set A to(at, in, within) B: A 를 B 로 선택하라, ~을 ~로 만들다, ~내로 조절하라

Set the electrical power source to the engine generators.
Do these steps to set the airplane in the takeoff configuration.
Set the stabilizer within 1 unit center of the green band.

99. should be ~: 반드시 ~되어야 한다

Hydraulic fluid found in the torque box should be cleaned.

100. show: 보여주다, 나타내다, 알려주다

The sound of the click shows that the actuator is set.

101. ~ shown in ~: ~에 보여주다

Perform a detailed visual inspection of the pulleys for conditions shown in (Figure 2).

102. shut off: 작동을 멈추다, 꺼지다

If the supply fans do not shut off when commanded,

103. skip ~: ~을 건너 뛰다, ~을 생략하다

To make the test easier, you may skip the following step and perform the voltage check.

104. speak into ~: ~에 대고 말하다

Speak into the captain's and first officer's boom microphone.

105. speak to ~: ~에 문의하다

If the dents are out of these limits, speak to the manufacturer.

106. stands for ~: ~을 말(의미)한다

Tool part numbers that are replaced or non-procurable are preceded by "Opt:", which stands for Optional.

107. such as ~: ~와 같은

Check the EWIS and the area around them for combustible material such as dust, lint and other surface contamination.

108. test ~: ~을 시험하다

To test the right system, put the ATC select switch to the No. 2 position.

109. to prevent ~: ~을 방지하기 위해, ~을 방지하려면

Secure the main landing gear trucks, if required, by rope to prevent rotation during the jacking operation.

110. to keep A in position: A를 제자리에 잡아주기 위해

Make sure that the spring clips apply pressure to keep the filter in position.

111. turn ~: ~을 돌리다

Do not turn the control wheel more than 100 degrees from the neutral position.

112. by use of ~: ~을 사용하여

Tests may be reviewed or run individually by use of the DATA and SELECT keys.

113. until ~: ~할 때까지

Set the AUTO/TEST/MANUAL switch to the MANUAL position until Mode S Address screen shows.

114. verify ~: ~을 확인하다, ~을 입증하다

Verify that the STBY RUD ON light on the overhead panel P5-3 is illuminated.

115. when ~: ~한 경우에, ~할 때에

Nose landing gear will move when the rudder pedals are moved.

116. while ~: ~를 하면서, ~하는 동안에

Increase the altitude on the cabin altitude warning switch, S128, while you monitor the altimeter.

117. with ~: ~을 이용하여, ~한 상태로

The center lower display UNIT is installed with the latch mechanism at the top.

118. within ~: ~내에서는, ~ 내에서

The minimum cable clearance from other parts is 0.20 inches, except 0.10 inches **within** 10 inches of a pulley or quadrant.

119. with incorporation of ~: ~을 수행한, ~이 수행된

Applicable to airplanes line number 596; and 1268 and on; and L/N 1-595 and 597-1267 **with incorporation of** SB 737-27-1253.

Appendix 3

Abbreviations & Acronyms used in Aircraft Maintenance

A

• A	Amber
• AACU	Antiskid/Autobrake Control Unit
• AAD	Assigned Altitude Deviation
• AAVM	Advanced Airborne Vibration Monitoring
• AAU	Audio Accessory Unit
• A/B	Autobrake
• A/C	Aircraft
• A/C	Air Conditioning
• AC	Alternating Current
• AC	Advisory Circular
• ACARS	Aircraft Communications Addressing and Reporting System
• ACAU	Air Conditioning Accessory Unit
• ACCUM	Accumulator
• ACCY	Accessory
• ACE	Accumulator Control Electronics
• ACESS	Advance Cabin Entertainment and Service System
• ACFT	Aircraft
• ACM	Air Cycle Machine
• ACMP	Alternating Current Motor Pump (See also EMP)
• ACMS	Aircraft Condition Monitoring System
• ACOC	Air Cooled Oil Cooler
• ACP	Audio Control Panel
• ACS	Air Conditioning System
• ACT	Active
• ACTR	Actuator
• A/D	Analog-to-Digital
• AD	Airworthiness Directives
• ADC	Air Data Computer
• ADF	Automatic Direction Finder
• ADI	Attitude Direction Indicator
• ADIRS	Air Data Inertial Reference System
• ADIRU	Air Data Inertial Reference Unit
• ADL	Airborne Data Loader
• ADM	Air Data Module
• ADP	Air Driven Pump
• ADR	Air Data Reference
• ADRS	Address
• ADS	Air Data System
• ADU	Air Drive Unit
• ADV	Advisory, Advance
• AEC	Aft Equipment Center
• AEM	Audio Entertainment Multiplexer
• AEP	Audio Entertainment Player
• AEVM	Advanced Engine Vibration Monitoring
• AFCS	Automatic Flight Control System
• AFDC	Air Flight Data Control
• AFDS	Autopilot Flight Director System
• AFL	Air Flow
• AFM	Airplane Flight Manual
• AFS	Auto Flight System
• AFT	Afterward
• AGB	Accessory Gearbox

- **AGCU** APU (Auxiliary) Generator Control Unit
- **AGL** Above Ground Level
- **AGS** Air/Ground System
- **AH** Alert Height
- **AHM** Aircraft Health Management
- **AI** Attitude Indicator
- **AI** Anti-Icing
- **AID** Aircraft Installed (Installation) Delay
- **AIDS** Airborne Integrated Data System
- **AIL** Aileron
- **AIM** Aircraft Identification Module
- **AIMS** Aircraft Information Management System
- **ALF** Aft Looking Forward
- **ALRT** Alert
- **ALS** Ambient Light Sensor
- **ALT** Altimeter
- **ALT** Altitude
- **ALT** Alternate
- **ALT HOLD** Altitude Hold
- **ALTM** Altimeter
- **ALTN** Alternate
- **AM** Amplitude Modulation
- **AMB** Ambient
- **AME** Amplitude Modulation Equivalent
- **AME** Aircraft Maintenance Engineer
- **AMM** Aircraft Maintenance Manual
- **AMO** Approved Maintenance Organization
- **AMOC** Approved Means of Compliance
- **AMP** Amplifier
- **AMP** Ampere
- **AMU** Audio Management Unit
- **AMUX** Audio Multiplexer
- **ANCMT** Announcement
- **ANCPT** Anticipate
- **ANCPTR** Anticipator
- **ANNC** Announcement
- **ANNUN** Annunciator
- **ANS** Ambient Noise Sensor
- **ANT** Antenna
- **ANTI-COLL** Anti-Collision
- **AOA** Angle Of Airflow
- **AOA** Angle Of Attack
- **AOC** Approach-On-Course
- **AOC** Air Oil Cooler
- **AOC** Airline Operations Control
- **AOG** Aircraft (Airplane) On Ground
- **A/P** Airplane
- **A/P** Autopilot
- **APB** Breaker
- **APB** Auxiliary APU Power Breaker
- **APCH** Approach
- **APID** Airplane Identification
- **APL** Airplane
- **APM** Airplane Personality Module
- **APP** Approach
- **APPL** Application
- **APPROX** Approximately
- **APS** APU Power Switch
- **APU** Auxiliary Power Unit
- **APUC** Auxiliary Power Unit Controller
- **AR, A/R** As Required
- **AR, A/R** Altitude Rate
- **ARINC** Aeronautical Radio Incorporated
- **ARPT** Airport
- **ARR** Arrival
- **A/S** Airspeed
- **ASA** Autoflight (Auto Land) Status Annunciator
- **ASC** Automatic Sensitive Control
- **ASCII** American Standard Code for Information Interchange
- **ASCPC** Air Supply Cabin Pressure Controller
- **ASCTS** Air Supply Control and Test System
- **ASCTU** Air Supply Control and Test Unit
- **ASE** Altimeter System Error
- **ASI** Airspeed Indicator
- **ASIC** Application Specific Integrated Circuit
- **ASG** APU starter-generator
- **ASM** Auto Throttle Servo Motor
- **ASM** Air Separation Module

- **ASP** Audio Select Panel
- **ASSY** Assembly
- **ASU** Attendant Service Unit
- **ASU** Aft Attendant Station
- **ASYM** Asymmetry
- **A/T** Auto Throttle
- **ATA** Air Transport Association
- **ATA** Actual Time of Arrival
- **ATB** Air Turn Back
- **ATC** Air Traffic Control
- **ATCRBS** Air Traffic Control Radar Beacon System
- **ATE** Automatic Test Equipment
- **ATIS** Air Terminal Information Service
- **ATE** Automatic Test Equipment
- **ATM** Atmosphere
- **ATR** Austin Trumbull Radio
- **ATT** Attitude
- **ATTD** Attendant
- **AUD** Audio
- **AUTH** Authority
- **AUTO** Automatic
- **AUX** Auxiliary
- **AVAIL** Available
- **AVC** Automatic Volume Control
- **AVM** Airborne Vibration Monitoring
- **AVN** Aviation
- **AWL** Airworthiness Limitation
- **AWI** Airworthiness Limitation Instructions
- **AZ** Azimuth

B

- **B** Both
- **BAL** Balance
- **BAR** Bleed Air Regulator
- **BARO** Barometric
- **BAT** Battery
- **BATT** Battery
- **BAV** Bleed Air Valve
- **BBL** Body Buttock Line
- **BC** Broadcast
- **BCF** Boeing Converted Freighter
- **BCN** Beacon
- **BCS** Block Check Sequence
- **BCU** Bus Control Unit
- **B/CRS** Back Course
- **BDCO** Boeing Designated Compliance Organization
- **BDU** BITE display unit
- **BDY BLK** Burndy Block
- **BFE** Buyer Furnished Equipment
- **BFO** Beat Frequency Oscillator
- **BGM** Boarding Music
- **BIT** Binary Digit
- **BIT** Built In Test
- **BITE** Built In Test Equipment
- **BK** Brake
- **BKGRD** Background
- **BKR** Breaker
- **BL** Buttock Line
- **BLD** Bleed
- **BLS** Bezel Light Sensor
- **BLW** Below
- **BMS** Boeing material specification
- **BMV** Brake Metering Valve
- **BOT** Bottom
- **BOV** Bias Out of View
- **BOW** Basic Operational Weight
- **BPCU** Bus Power Control Unit
- **BPI** Brake Pressure Indicator
- **BPT** Break Power Transfer
- **BRG** Bearing
- **BRIL** Brilliance
- **BRK** Brake
- **BRT** Bright
- **BRT** Brightness
- **BS** Body Station
- **BSC** Basic
- **BSCU** Brake System Control Unit
- **BSI** Borescope Inspection
- **BST** Boost
- **BSV** Burner Staging Valve
- **BT** Bus Tie
- **BTB** Bus Tie Breaker
- **BTB** Block Turn Back
- **BTL** Bottle
- **BTLCS** Brake Torque Limiting Control System

- BTMU Brake Temperature Monitor Unit
- BTU British Thermal Unit
- BVCU Bleed Valve Control Unit
- BYP Bypass

C

- C Celsius
- C Cold
- CACTS Cabin Air Conditioning & Temperature Control System
- CAA Civil Aviation Authority
- CAB Cabin
- CACP Cabin Area Control Panel
- CAD Combiner Alignment Detector
- CADC Central Air Data Computer
- CADS Central Air Data System
- CAI Cabin Altitude Indicator
- CAL Calibrate
- CALC Calculate
- CANC, CANX Cancel
- CAP Contact Authorized Proposal, Capture
- CALIB Calibrator
- CAPT Captain
- CAR Cargo
- CAS Computed Airspeed
- CAT Category
- CAUT Caution
- C/B Circuit Breaker(s)
- CBAL Counterbalance
- CBT Computer Based Training (Teaching)
- CC Cubic Centimeter(s)
- CCA Circuit Card Assembly
- CCDL Cross Channel Data Link
- CCL Cargo Control Logic
- CCM Cargo Control Module
- CCN Compatibility Class Number
- CCP Camera Control Panel
- CCU Camera Control Unit
- CCU Cargo Control Unit
- CCW Counterclockwise
- CD Case Drain
- CDCCL Critical Design Configuration Control Limitations
- CDI Course Deviation Indicator
- CDL Configuration Deviation List
- CDP Compressor Discharge Pressure
- CDS Common Display System
- CDU Control Display Unit
- CEC Center Equipment Center
- CFDS Centralized Fault Detection System
- CFE Customer Furnished Equipment
- CFEC Center Fuselage Equipment Center
- CFM Cubic Feet per Meter
- CFR Code of Federal Regulations
- CFS Cubic Feet per Second
- CG Center of Gravity
- CGO Cargo
- CH Channel
- CHAN Channel
- CHAP Chapter
- CHAR Character
- CHG Change
- CHGR Charger
- CHKPT Checkpoint
- CHP Common HUD Provisioned (Aircraft)
- CHRGR Charger
- CHSP Course Heading Select Panel
- CIC Cabin Interphone Controller
- CIH Cabin Interphone Handset
- CIRC Circulate
- CIS Cabin Interphone System
- CIT Compressor Inlet Temperature
- CIWS Central Instrument Warning System
- CKT Circuit
- CL Center Lower
- CL Center Line
- CLASS Classification
- CLB Climb
- CLCP Central Lateral Control Package
- CLG Ceiling

• CLK	Clock
• CLNG	Cooling
• CLP	Control Law Processor
• CLR	Clear
• CLS	Continuous Level Sensor
• CLS	Cabin Lighting System
• CLSD	Closed
• CM	Centimeter
• CM	Configuration Module
• CMC	Central Maintenance Computer
• CMCS	Central Maintenance Computer System
• CMD	Command
• CMM	Component Maintenance Manual
• CMP	Compare
• CMP	Configuration Maintenance Procedures
• CMPS	Centimeters Per Seconds
• CMPTR	Computer
• CMR	Certification Maintenance Requirements
• CMS	Central Maintenance System
• CMS	Cabin Management System
• CMU	Communications Management unit
• CNTRL	Control
• CNX	Cancelled
• C/O	Change Over
• COAX	Coaxial
• COC	Customer Originated Change
• COE	Combiner Optical Element
• COEF	Coefficient
• COF MKR	Coffee Maker
• COL	Column
• COLL	Collision
• COM	Common
• COMAT	Company Manual/Company Material
• COMB	Combination
• COM/NAV	Communication/ Navigation
• COMND	Command
• COMP	Compressor
• COMPT	Compartment
• COMSAT	Communications Satellite
• CON	Continuous
• COND	Condition
• COND	Conditioning
• CONFIG	Configuration
• CONN	Connector
• CONST	Constant
• CONT	Control
• CONT	Controller
• CONT PNL	Control Panel
• CONV	Converter
• CONVERG	Convergence
• COOL	Coolant
• COR	Corrector
• CORR	Correction
• CNTL	Control
• CNTRL	Control
• COMM	Communication
• COWL	Cowling
• CP	Control Panel
• CP	Center Of Pressure
• CPC	Cabin Pressure Controller
• CPCP	Corrosion Prevention & Control System
• CPCS	Cabin Pressure Control System
• CPCU	Cabin Pressure Control Unit
• CPDLC	Controller Pilot Data Link Communication
• CPM	Cycles Per Minute
• CPS	Cycles Per Second
• CPU	Central Processing Unit
• CRIT	Critical
• CRKG	Cranking
• CRS	Course
• CRT	Cathode Ray Tube
• CRZ	Cruise
• CSB	Compressor Stability Bleed
• CSD	Constant Speed Drive
• CSEU	Control System Electronic Unit
• CSL	Console
• CSLSV	Cycles Since Last Shop Visit
• CSMU	Crash Survivable Memory Unit
• CSN	Cycles Since New
• CSO	Cycles Since Overhaul

- **CSPR** Cycles Since Performance Restoration
- **CT** Current Transformer
- **CT** Control Transformer
- **CTA** Current Transformer Assembly
- **CTAI** Cowl Thermal Anti-Ice
- **CTC** Cabin Temperature Controller
- **CTG** Cycles To Go
- **CTR** Center
- **CTS** Conversational Terminal System
- **CTRL** Control
- **CU** Center Upper
- **CU** Control Unit
- **CURR** Current
- **CV** Check Valve
- **CVR** Cockpit Voice Recorder
- **CW** Continuous Wave
- **C/W** Control Wheel
- **C/W** Clockwise
- **CWS** Control Wheel Steering
- **CWT** Center Wing Tank
- **CYL** Cylinder

D

- **DA** Drift Angle
- **D/A** Digital to Analog
- **DAA** Digital/Analog Adapter
- **DABS** Discrete Addressable Beacon System
- **DAC** Digital Audio Control
- **DADC** Digital Air Data Computer
- **DAR** Digital Aids Recorder
- **DAR** Designated Airworthiness Representative
- **DATR** Digital Audio Tape Reproducer
- **DB** Database
- **DB** Decibel
- **DC** Direct Current
- **DC** Daily Check
- **DCCS** Direct Current Sensor
- **DCDR** Decoder
- **DCIR** Direct Current Isolation Relay
- **DCTCU** Direct Current Tie Control Unit
- **DCV** Directional Control Valve
- **DDG** Dispatch Deviations Guide
- **DEC** Decimal
- **DECR** Decrease
- **DED** Dead Ended Shield
- **DEF** Definition
- **DEFL** Deflection
- **DEG** Degree
- **DEL** Deleted
- **DEL** Diagram Equipment List
- **DEM** Demand
- **DEN** Density
- **DE-MUX** De-multiplexer
- **DEP** Departure
- **DEP** Depressurize
- **DEPRESS** Depressurization
- **DEPT** Departure
- **DEPT** Department
- **DES** Descent
- **DES** Designated Engineering Representative
- **DEST** Destination
- **DET** Detector
- **DEU** Display Electronics Unit
- **DEV** Deviation
- **DEVN** Deviation
- **DF** Direction Finder
- **DFCS** Digital Flight Control System
- **DFDAC** Digital Flight Data Acquisition Card
- **DFDAU** Digital Flight Data Acquisition Unit
- **DFDMU** Digital Flight Data Management Unit
- **DFDR** Digital Flight Data Recorder
- **DH** Decision Height
- **DI** De-Icing
- **DIFF** Differential
- **DES** Designated Engineering Representative
- **DIG** Digital
- **DIGTL** Digital
- **DIM** Dimmer
- **DIP** Dual In-Line Package
- **DIR** Direct
- **DIR** Director

- **DIR** Direction
- **DIS** Discrete
- **DISASSY** Disassembly
- **DISC** Disconnect
- **DISCH** Discharge
- **DISEN** Disengage
- **DISP** Dispatch
- **DISPL** Displacement
- **DIST** Distance
- **DITS** Digital Information Transfer System
- **DIU** Digital Interface Unit
- **DK** Deck
- **DLY** Delay
- **DMA** Direct Memory Access
- **DME** Distance Measuring Equipment
- **DMI** Deferred Maintenance Item
- **DMM** Data Memory Module
- **DMP** Data Management Processor
- **DMPT** Damper
- **DMS** Digital Media Server
- **DMU** Data Management Unit
- **DN** Down
- **DOC** Document
- **DOT** Department Of Transportation
- **DP** Differential Pressure
- **DP** Differential Protection
- **DPA** Digital Pre-Assembly
- **DPCT** Differential Protection Current Transformer
- **DPLY** Deploy
- **DR** Deploy
- **DSBL** Disable
- **DSM** Door(s)
- **DSP** Display Select Panel
- **DSPL** Display (Command)
- **DTD** Dated
- **DTG** Distance-To-Go
- **DTI** Damaged Tolerance Inspection
- **DU** Display Unit
- **DUDB** Display Unit Data Base
- **DUP** Duplicate
- **DV** Distinguished Visitor
- **DWG** Drawing
- **DYB** Dynamic Braking
- **DYN** Dynamic

E

- **E** East
- **E** Empty
- **E/A** Engineering Authorization
- **EA** Each
- **EADI** Electronic Attitude Director Indicator
- **EASA** European Aviation Safety Agency
- **EAU** Engine Accessory Unit
- **EBS** Engine Bleed System
- **EBU** Engine Build Up
- **ECC** Eccentric
- **ECM** Engine Condition Monitoring
- **ECON** Economy
- **ECS** Environmental Control System
- **ECU** Electronic Control Unit
- **ECU** Engine Control Unit
- **E/D** End of Descent
- **EDFCS** Enhanced Digital Flight Control System
- **EDIU** Engine Data Interface Unit
- **EDP** Engine-Driven Pump
- **EDSD** Electrical Discharge Sensitive Device
- **EE** Electronic Equipment
- **E/E** Electrical and Electronic
- **EEC** Electronic Equipment Compartment
- **EEC** Electronic Engine Control
- **EEL** Emergency Equipment List
- **EEPROM** Electrically Erasable Programmable Read Only Memory
- **EFCP** EFIS Control Panel
- **EFF** Effectivity
- **EFI** Electronic Flight Instrument
- **EFIS** Electronic Flight Instrument System
- **EFIS/CP** EFIS Control Panel
- **EFS** Elevator Feel Shift

- **EFSM** Elevator Feel Shift Module
- **EGPWS** Enhanced Ground Proximity Warning System
- **EGT** Exhaust Gas Temperature
- **EHSI** Electronic Horizontal Situation Indicator
- **EHSV** Electrohydraulic Servo Valve
- **EICAS** Engine Indicating and Crew Alerting System
- **EICAS/CP** EICAS Control Panel
- **EIS** Engine Indicating System
- **EIU** EFIS/EICAS Interface Unit
- **EL** Elevation
- **ELACT** Electronic Actuator
- **ELCCR** Electrical Liaison Change Commitment Record
- **ELCU** Electrical Load Control Unit
- **ELEC** Electrical
- **ELEV** Elevator
- **ELEX** Electronics
- **ELMS** Electrical Load Management System
- **ELT** Emergency Locator Transmitter
- **ELV** Elevation
- **EMAT** Electronic Maintenance Authoring Tool
- **EMC** Electromagnetic Compatibility
- **EMDP** Electric Motor-Driven Pump
- **EMERG** Emergency
- **EMP** Electric Motor Pump (See also ACMP)
- **EMP** Essential Maintenance Provider
- **ENBL** Enable
- **ENG** Engine
- **ENGA** Engage
- **ENT** Entertainment
- **ENT** Enter
- **ENTMT** Entertainment
- **ENWY** Entryway
- **EO** Engineering Order
- **E/O** Engine Out
- **EP** External Power
- **EPC** External Power Contactor
- **EPGS** Electrical Power Generating System
- **EPR** Engine Pressure Ratio
- **EPRL** Engine Pressure Ratio Limit
- **EPROM** Erasable Programmable Read Only Memory
- **EQL** Equalizer
- **EQPMT** Equipment
- **EQUIV** Equivalent
- **ERM** Engineering Repair Manual
- **ERP** Eye Reference Point
- **ESC** Escape
- **ESCC** Electrical Supply and Control Center
- **ESD** Electrostatic Sensitive Device
- **ESDS** Electrostatic Discharge Sensitive
- **ESI** Engine Start Inhibit
- **ESNTL** Essential
- **ESS** Essential
- **ET** Elapsed Time
- **ETA** Estimated Time of Arrival
- **ETC** Electronic Temperature Control
- **ETOPS** Extended Twin (Engine) Operations
- **ETR** Estimated Time of Release
- **EVAC** Evacuation
- **EVBC** Engine Vane and Bleed Control
- **EVSC** Engine Vibration Signal Conditioner
- **E-W** East-West
- **EWIS** Electrical Wiring Interconnection System
- **EX** Example
- **EXC** Excitation
- **EXCD** Exceedance
- **EXCH** Exchanger
- **EXCHR** Exchanger
- **EXCT** Excitation
- **EXCTR** Exciter
- **EXEC** Execute
- **EXT** External
- **EXTD** Extended
- **EXTIN** Extinguish
- **EXTIN** Extinguished

- **EXTING** Extinguishing
- **EZAP** Enhanced Zonal Analysis Procedure

F

- **F** Fahrenheit
- **F** Frequency
- **F** Filter
- **F** Full
- **FAA** Federal Aviation Administration
- **FAC** Final Approach Course
- **FADEC** Full Authority Digital Engine Control
- **FAEEC** Full Authority Electronic Engine Control (PW)
- **FAFC** Full Authority Fuel Control
- **FANS** Future Air Navigation System
- **FAR** Federal Aviation Regulations
- **F/A Ratio** Fuel/Air Ratio
- **FBW** Fly-by-Wire
- **FC** Flight Cycles
- **F/C** Flight Compartment
- **FCC** Flight Control Computer
- **FCES** Flight Control Electronic System
- **FCEU** Flight Control Electronic Unit
- **FCS** Flight Control System
- **FCSOV** Flow Control and Shutoff Valve
- **FCTN** Function
- **FCU** Flow Control Unit
- **FCU** Flap Control Unit
- **FCV** Flow Control Valve
- **F/D, FD, FLT DIR** Flight Director
- **FDAU** Flight Data Acquisition Unit
- **FDE** Flight Deck Effect
- **FDEVSS** Flight Deck Entry Video Surveillance System
- **FDH** Flight Deck Handset
- **FDR** Flight Data Recorder
- **FDRS** Flight Data Recorder System
- **FEXT** Fire Extinguisher
- **FF** Fuel Flow
- **FFCCV** Fan Frame Compressor Case Vertical (Sensor)
- **FFR** Fuel Flow Regulator (RR)
- **FGN** Foreign
- **FH** Flight Hour
- **FIM** Fault Isolation Manual
- **FJCC** Fuel Jettison Control Card
- **FL** Flight Level
- **FL CH** Flight Level Change
- **FLMTR** Flowmeter
- **FLS** Field Loadable Software
- **FLT** Flight
- **FLT ALT** Flight Altitude
- **FLT CONT** Flight Control
- **FLT DK** Flight Deck
- **FLTR** Filter
- **FM** Frequency Modulation
- **FMA** Flight Mode Annunciator/ Annunciation
- **FMC** Flight Management Computer
- **FMCS** Flight Management Computer System
- **FMS** Flight Management System
- **FMU** Fuel Metering Unit
- **FMV** Fuel Metering Valve
- **FO, F/O** First Officer
- **F/OBS** First Observer
- **FOC** Fuel/Oil Cooler
- **FP** Fuel Pressure
- **FPA** Flight Path Angle
- **FPAC** Flight Path Acceleration
- **FPM** Feet Per Minute
- **FPS** Feet Per Second
- **FPV** Flight Path Vector
- **FQIS** Fuel Quantity Indicating System
- **FQPU** Fuel Quantity Processor Unit
- **FQS** Fuel Quantity System
- **FREQ** Frequency
- **FRM** Fault Reporting Manual
- **FS** Fast Slow
- **FSB** Fasten Seat Belt
- **FSEIC** Fuel System EICAS Interface Card
- **FSEU** Flap/Slat Electronic Unit
- **FSK** Flight Spare Kit

- **FSMC** Fuel System Management Card
- **FSPM** Flap/Stabilizer Position Module
- **FSU** Fuel Summation Unit
- **FT** Foot/Feet
- **FT** Functional Test
- **FT-LB** Foot-Pound
- **FTR** Flight Time Recording
- **FTX** Fast Transmit
- **FU** Fuel Used
- **FUNC** Functional
- **FUS** Fuselage
- **FWD** Forward

G

- **G** Gravity (Acceleration)
- **GA** Gage
- **G/A** Go-Around
- **G/A** Ground-to-Air
- **GAL** Gallon
- **GALY** Galley
- **GB** Generator Breaker(s)
- **GBAS** Ground Based Augmentation System
- **GCB** Generator Control Breaker
- **GCCT** Generator Control Current Transformer
- **GCR** Generator Control Relay
- **GCS** Generator Control Switch
- **GCU** Generator Control Unit
- **GEN** Generator
- **GG** Graphics Generator
- **GG CCA** Graphics Generator Circuit Card Assembly
- **GHB** Ground Handling Bus
- **GHR** Ground Handling Relay
- **GLS** GNSS (GPS) Landing System
- **GMM** General Maintenance Manual
- **GMT** Greenwich Mean Time
- **GND** Ground
- **GND RET** Ground Return
- **GNSS** Global Navigation Satellite System
- **GOV** Governor
- **GP** General Purpose
- **G/P** Glide Path
- **GPH** Gallons Per Hour
- **GPM** Gallons Per Minute
- **GPIP** Glide Path Intercept Point
- **GPH** Gallons Per Hour
- **GPM** Gallons Per Minute
- **GPS** Global Positioning System
- **GPWC** Ground Proximity Warning Computer
- **GPWM** Ground Proximity Warning Module
- **GPWS** Ground Proximity Warning System
- **GR** Gear
- **GRD** Ground
- **GRWT** Gross Weight
- **G/S** Glide Slope
- **GS** Ground Speed
- **GSB** Ground Service Bus
- **GSE** ground support equipment
- **GSPR** Gasper
- **GSR** Ground Service Relay
- **GSSR** Ground Service Select Relay
- **GSTR** Ground Service Transfer Relay
- **G/T** Gear Train
- **GW** Gross Weight

H

- **H** Hour
- **H** Heater
- **H** Hot
- **HI** High
- **HAP** HGS Annunciator Panel
- **HAZMAT** Hazardous Material
- **HC** HGS Computer
- **HCP** HGS Control Panel
- **HDG** Heading
- **HDG HLD** Heading Hold
- **HDG SEL** Heading Select
- **HDLG** Handling
- **HDS** Horizontal Drive Shaft

- **HEX** Hexadecimal
- **HF** High Frequency
- **HG** Mercury
- **HGS** Heads Up Guidance System
- **HHDU** Hand Held Data Unit
- **HIRF** High Intensity Radiation Field
- **HIRF** High Intensity Radiation Field
- **HIV** Hydraulic Isolation Valve
- **HLCU** High Lift Control Unit
- **HLD** Hold
- **HMU** Hydro-Mechanical Unit
- **HND** Hand
- **HNDBK** Handbook
- **HOT** High Oil Temperature
- **HOR/HORIZ** Horizontal
- **HP** Horse Power
- **HP** High Pressure
- **hPA** Hecto Pascals
- **HPBV** High Pressure Bleed Valve
- **HPC** High Pressure Compressor
- **HPI** High Pressure Indicator
- **HPSOV** High Pressure Shut Off Valve
- **HPT** High Pressure Turbine
- **HPTACC** High Pressure Turbine Active Clearance Control
- **HPWS** High Pressure Water Separator
- **HQI** Hydraulic Quantity Indicator
- **HR** Hour
- **HRD** High Rate Discharge
- **HRDW** Hardware
- **HSBL** Horizontal Stabilizer Buttock Line
- **HSI** Horizontal Situation Indication
- **HST** Horizontal Stabilizer Tank
- **HT** Heat
- **HT** Height
- **HUD** Head Up Display
- **HV** High Voltage
- **HVPS** High Voltage Power Supply
- **H/W** Hardware
- **H/WIND** Head Wind
- **HYD** Hydraulic
- **HYDIM** Hydraulic Interface Module
- **HYQUIM** Hydraulic Quantity Interface Module
- **Hz** Hertz (Cycles per Second)

I

- **IAN** Integrated Approach Navigation
- **IAS** Indicated Airspeed
- **IBIT** Initiated Built In Test
- **IBVSU** Instrument Bus Voltage Sense Unit
- **IC, I/C** Integrated Circuit
- **I/C** Inspection/Check
- **I/C** Interphone Communication
- **ICA** Instruction for Continued Airworthiness
- **ICAO** International Civil Aviation Organization
- **ICU** Instrument Comparator Unit
- **ICU** Interface Control Unit
- **ID** Identification
- **IDENT** Identification
- **IDENT** Identity
- **IDG** Integrated Drive Generator
- **IDS** Integrated Display System
- **IDU** Interactive Display Unit
- **IF** Intermediate Frequency
- **IFR** Instrument Flight Rules
- **IFSAU** Integrated Flight System Accessory Unit
- **IFSD** In Flight Shut Down
- **IGB** Inlet Gear Box
- **IGN** Ignition
- **IGS** Integrated Graphic System
- **IGV** Inlet Guide Vane
- **IGVA** Inlet Guide Vane Actuator
- **IHC** Integrated Handset Controller
- **ILES** Inboard Leading Edge Station
- **ILLUM** Illuminate
- **ILLUS** Illustrate
- **ILLUS** Illustration
- **ILS** Instrument Landing System
- **IM** Inner Marker
- **IMC** Instrument Meteorological Conditions
- **IN** Inch
- **INBD** Inboard
- **INCR** Increase
- **IND** Indication
- **IND** Indicator

- INFLT In-Flight
- INFO Information
- INHB Inhibit
- IN HG Inches of Mercury
- IN HG Inches of Mercury
- INIT Initialize
- INIT Initialization
- INIT/REF Initialization Reference
- INOP Inoperative
- INOP Inoperable
- INP Input
- INPH Interphone
- INS Inertial Navigation System
- INSP Inspection
- INSP Inspector
- INST Instrument
- INSTL Installation
- INSTR Instrument
- INT Interphone
- INT Interrogator
- INT Internal
- INT Integrating
- INTAM Information To Airmen
- INTC Intercept
- INTCHG Interchangeable
- INTCON Interconnection
- INTER Interrupt
- INTERCOM Intercommunication
- INTFC Interface
- INTK Intake
- INTLK Interlock
- INTMT Intermittent
- INTPH Interphone
- INU Inertial Navigation Unit
- INV Inverter
- I/O Input/Output
- IOC Input Output Controller
- IOEU Inboard Overhead Electronics Unit
- IOP Input/Output Processor
- IOS Input Output Subsystem
- IP Intermediate Pressure
- IPC(M) Illustrated Parts Catalog (Manual)
- IPM Inspection Program Manual
- IPL Illustrated Parts List
- IPS Inches Per Second
- IR Infra-Red
- IR Inertial Reference
- IRMP Inertial Reference Mode Panel
- IRS Inertial Reference System
- IRU Inertial Reference Unit
- ISA International Standard Atmosphere
- ISB Intersystem Bus
- ISDR Internet Service Difficulty Reporting
- ISDU Inertial System Display Unit
- ISFD Integrated Standby Flight Display
- ISIN Isolation
- ISO International Standards Organization
- ISO Valve Isolation Valve
- ISPS In-Seat Power Supply
- ISSS Instrument Source Select Switch
- ISV Isolation Valve
- IVS Inertial Vertical Speed
- IU Index Units
- IUL In-Use Light

J

- J Junction
- JAM Jammed
- JAM Jamming
- J-BOX Junction Box
- JTSN, JETT Jettison

K

- K Knot
- K Kilo
- KG Kilogram
- KHz Kilo Hertz
- KIAS Knots Indicated Airspeed
- KM Kilometer
- KT, KTS Knots (Nautical Miles Per Hour)

- **KVA** Kilovolt-Ampere
- **KVAR** Kilovolt-Ampere Reactive
- **KW** Kilowatts
- **KWH** Kilowatts-Hour
- **KWR** Kilowatts Reactive

L

- **L** Left
- **LA** Lightning Arrester
- **LAC** Local Area Controller
- **LAM** Laminate
- **LAND ALT** Landing Altitude
- **LAS** Load Alleviation System
- **LAT** Latitude
- **LAT** Lateral
- **LAT DEV** Lateral Deviation
- **LAV** Lavatory
- **LBL** Left Buttock Line
- **LBS** Pounds
- **LC** Lower Center
- **LCA** Lateral Control Actuator
- **LCD** Liquid Crystal Display
- **LCF** Large Cargo Freighter
- **LCL** Local
- **LCM** Logic Control Module
- **LD** Load
- **LE** Leading Edge
- **LED** Light Emitting Diode
- **LED** Leading Edge Device
- **LF** Left Front
- **LF** Low Frequency
- **LH** Left Hand
- **LIB** Left Inboard
- **LIM** Limit
- **LIM SW** Limit Switch
- **LDG** Landing Gross Weight
- **LG** Landing Gear
- **LGW** Landing Gross Weight
- **LGHTNG** Lightning
- **LKD** Locked
- **LLP** Life Limited Parts
- **LMP** Lamp
- **LNAV** Lateral Navigation
- **LO** Low
- **LO** Lock Out
- **LOB** Left Outboard
- **LOC** Localizer
- **LOC** Local
- **LOM** Localizer Outer Mark
- **LONG** Longitude
- **LOP** Low Oil Pressure
- **LP** Low Pressure
- **LP** Lightning Protector
- **LPC** Low Pressure Compressor
- **LPM** Liters Per Minute
- **LPT** Low Pressure Turbine
- **LPTACC** Low Pressure Turbine Active Clearance Control
- **LRC** Long Range Cruise
- **LRD** Low Rate Discharge
- **LRRA** Low Range Radio Altimeter
- **LRU** Line Replaceable(Replacement) Unit
- **LSDA** Low Speed Digital To Analog
- **LSK** Line Select Key
- **LSU** Lavatory Service Unit
- **LT** Light
- **LTD** Limited
- **LTG** Lighting
- **LVDT** Linear Variable Differential Transformer
- **LVL** Level
- **LVL CHG** Level Change
- **LVR** Lever
- **LW** Left Wing
- **LWR** Lower

M

- **M** Month
- **M** Minute
- **M** Meters
- **M** Milli
- **M** Mach
- **MA** Master
- **MAC** Mean Aerodynamic Chord
- **MACH** Mach number
- **MAG** Magnetic

- **MAI** Multiplexer Action Item
- **MAINT** Maintenance
- **MAN** Manual
- **MAPA** Maintenance And Pool Agreement
- **MASI** Mach Airspeed Indicator
- **MAWEA** Modularized Avionics and Warning Electronics Assembly
- **MAX** Maximum
- **MB** Marker Beacon
- **MB** Millibars
- **MBR** Main Battery Relay
- **MC** Master Change
- **MCC** Maintenance Control Center
- **MCDP** Maintenance Control and Display Panel
- **MCDU** Multi-Function Control Display Unit
- **MCI** Maintenance Continued Item
- **MCP** Mode Control Panel (Autopilot/Autothrottle)
- **MCS** Master Control Switch (APU)
- **MCU** Modular Concept Unit, Master Control Unit
- **MDA** Minimum Descent Altitude
- **MDDR** Maintenance Defect Deferral Report
- **MDL** Module
- **MD & T** Master Dim and Test
- **ME** Mechanical Equipment
- **MEDB** Model/Engine Data Base
- **MEDCOM** Medical Communications
- **MEC** Maintenance Equipment Center
- **MEC** Main Engine Control
- **MECH** Mechanical
- **MECH** Mechanic
- **MED** Main Entry Door
- **MEL** Minimum Equipment List
- **MEM** Memory
- **MES** Main Engine Start
- **MEW** Manufacture's Empty Weight
- **MFD** Multi-Function Display
- **MFD** Manufactured
- **MG** Motor Generator
- **MGT** Management
- **MHRS** Magnetic Heading Reference System
- **MHz** Mega Hertz
- **MI** Miles
- **MIC** Microphone
- **MICRO-P** Microprocessor
- **MID** Middle
- **MIDU** Multipurpose Interactive Display Unit
- **MIN** Minute
- **MIN** Minimum
- **MIP** Maintenance and Inspection Program
- **MIS** Maintenance Interruption Summary
- **MISC** Miscellaneous
- **MISR** Maintenance Interruption Summary Report
- **MKR BCN** Marker Beacon
- **MLG** Main Landing Gear
- **MLS** Microwave Landing System
- **MM** Manufacturer's Maintenance Manual
- **MM** Middle Marker
- **MM** Millimeter
- **MMEL** Master Minimum Equipment List
- **MMO** Mach Maximum Operating
- **MMR** Multimode Receiver
- **MMUX** Main Multiplexer
- **MN** Main
- **MNFST** Manifest
- **MNVR** Maneuver
- **MO** Month
- **MO** Modification Order
- **MOD** Modification
- **MOD** Modified
- **MOD** Module
- **MOE** Maintenance Organization Exposition
- **MON** Monitor
- **MOSFET** Metallic Oxide Semiconductor Field Effect Transistor
- **MP** (Approved Contract) Maintenance Provider
- **MP** Microprocessor

- **MPC** Maintenance Program Change
- **MPD** Maintenance Planning Data Document
- **MPG** Miles Per Gallon
- **MPH** Miles Per Hour
- **MPID** Maintenance Planning Information Document
- **MR** Modification Revision
- **MRB** Maintenance Review Board
- **MRI** Maintenance Requirement Item
- **MRK** Marker
- **MRO** Maintenance Repair and Overhaul
- **MRR** Maintenance Reliability Report
- **MSEC** Milli-Second
- **MSG** Message
- **MSI** Maintenance Significant Item
- **MSL** Mean Sea Level
- **MST** Master
- **MSU** Mode Select Unit
- **MTBF** Mean Time Between Failure
- **MTBR** Mean Time Between Removal
- **MTC** Maintenance
- **MTCHG** Matching
- **MTG** Muting
- **MTG** Miles To Go
- **MTOW** Maximum Take-Off Weight
- **MTR** Magnetic Track
- **MTRS** Meters
- **MTW** Maximum Taxi Weight
- **MU** Management Unit
- **MUX** Multiplex
- **MV** Milli Volts
- **MW** Master Warning
- **mW** Milli Watt
- **MX** Maintenance
- **MZFW** Maximum Zero Fuel Weight

N

- **N** Normal
- **N** North
- **NA, N/A** Not Applicable
- **NAC** Nacelle
- **NAV** Navigation
- **NAV RAD** Navigation Radio
- **NC** Not Connected
- **NCD** No Computed Data
- **NCT** Neutral Current Transformer
- **ND** Navigational Display
- **NDB** Navigation Database
- **NDI** Non-Destructive Inspection
- **NDT** Non-Destructive Testing
- **NEA** Nitrogen Enriched Air
- **NEADS** Nitrogen Enriched Air Distribution System
- **NEG** Negative
- **NGS** Nitrogen Generation System
- **NIS** Not In Stock
- **NLG** Nose Landing Gear
- **NM** Nautical Miles
- **NN** A Number From 01 To 99
- **No.** Number
- **NON STD** Non Standard
- **NORM** Normal
- **NOTAM** Notice To Airmen
- **NPRM** Notice of Proposed Rule Making
- **NPS** Navigation Performance Scales
- **NR, N/R** Not Required
- **N-S** North-South
- **NSN** No Serial Number
- **NSS** Neutral Shift Sensor
- **NV** Non-Volatile
- **NVM** Non-Volatile Memory
- **NWS** Nose Wheel Steering
- **NWW** Nose Wheel Well
- **NVM** Non-Volatile Memory

O

- **OAP** Output Audio Processor
- **OAT** Outside Air Temperature
- **OBS** Observer
- **OC** On Course
- **OC** On Condition
- **O/D** Out Of Detent

- **ODA** Organization Designation Authorization
- **OEA** Oxygen Enriched Air
- **OEM** Original Equipment Manufacturer
- **OEV** Overboard Exhaust Valve
- **OEW** Operating Empty Weight
- **OFP** Operational Flight Program
- **OFL** Outflow
- **OFST** Offset
- **OHU** Overhead Unit
- **OK** Good or Pass
- **OLS** Oil Level Sensor
- **OM** Outer Marker
- **OM** Outboard Maintenance System
- **OOEU** Outboard Overhead Electronics Unit
- **OOOI** Out, Off, On, In
- **OPAS** Overhead Panel ARINC 629 System
- **OPBC** Overhead Panel Bus Controller
- **OPC** Operational Program Configuration
- **OPP** Opposite
- **OPR** Operator
- **OPR** Operate
- **OPS** Operational Software
- **OPT** Optical
- **OPTL** Optional
- **OSHA** Occupational Safety and Health Administration
- **OSC** Oscillator
- **OSS** Over Station Sensor
- **OTSOV** Over Temperature Shutoff Valve
- **OU** Outlet Unit
- **OUTBD** Outboard
- **OVDR** Over door
- **OVFL** Overfill
- **OVFL** Overflow
- **OVHD** Overhead
- **OVHT** Overheat
- **OVRD** Override
- **OVSP** Overspeed
- **OVWG** Over Wing
- **OXY, Oz** Oxygen

P

- **P** Push
- **P** Pressure
- **P** Panel
- **P** Pitch
- **Po** Aircraft Static Air Pressure
- **P□** Inlet Pressure
- **PA** Passenger Address
- **PA** Public Address
- **PA** Power Amplifier
- **PA/CI** Passenger Address/Cabin Interphone
- **PAI** Principal Avionics Inspector
- **PALCS** Passenger Address Level Control Sensor
- **PAM** Performance Assessment Monitor
- **PAMB** Ambient Pressure
- **PARAMS** Parameters
- **PAS** Passenger Address System
- **PASS** Passenger
- **PB** Push-Button
- **PBOM** Partial Bill Of Material
- **PCMCIA** Personal Computer Memory Card International Association
- **PC** Patch
- **PC** Printed Circuit
- **PCA** Power Control Actuator
- **PCP** Power Control Package
- **PCP** Pilot Call Panel
- **PCT** Percent
- **PCU** Power Control Unit
- **PCU** Passenger Control Unit
- **PCW** Previously Complied With
- **P&B** Pressurizing and Drain
- **PDB** Performance Data Base
- **PDBO** Pressure Deceleration Bleed Override
- **PDDB** Performance Default Data Base
- **PDL** Portable Data Loader

- **PDP** Power Distribution Panel
- **PDR** Publication Data Request
- **PDSC** Pre-Departure Service Check
- **PDU** Power Drive Unit
- **PED** Personal Electronic Device
- **PERF** Performance
- **PERF FACT** Performance Factor
- **PERF INT** Performance Initialization
- **PERS** Personnel
- **PES** Passenger Entertainment System
- **PF** Power Factor
- **PFC** Primary Flight Computer
- **PFCS** Primary Flight Control System
- **PFD** Primary Flight Display
- **PFIDS** Passenger Flight Information Display System
- **PGM** Program
- **PH** Phase
- **PIREPS** Pilots Reports
- **PIS** Passenger Information Sign
- **PJTR** Projector
- **PK** Pack
- **PK** Park
- **PKG** Packing
- **PKG** Package
- **PLA** Programmed Logic Array
- **PLF** Present Leg Faults
- **PLI** Pitch Limit Indication(Indicator)
- **PLL** Phase Lock-Loop
- **PLN** Plan
- **PLS** Point Level Sensor
- **PM** Phase Modulation
- **PMA** Permanent Magnet Alternator
- **PMA** Portable Maintenance Aids
- **PM-CPDLC** Protected Mode Controller Pilot Data Link Communication
- **PMG** Permanent Magnet Generator
- **PMI** Principal Maintenance Inspector
- **PMS** Performance Management System
- **P/N** Part Number
- **PNEU** Pneumatic
- **PNL** Panel
- **PO** Outside Air Pressure
- **POI** Principal Operations Inspector
- **POR** Power On Reset
- **POS** Position
- **POS** Positive
- **POS INT** Position Initialization
- **POSN** Position
- **POS REF** Position Reference
- **POT W** Potable Water
- **PP** Power Plant
- **PPH** Pounds Per Hour
- **PPIU** Programming and Position Interface Unit
- **PPM** Pounds Per Minute
- **PPOS** Present Position
- **PPR** Paper
- **PPS** Precision Positioning Service
- **PR** Pair
- **P/R** Push To Reset
- **PRAM** Pre-Recorded Announcement Machine
- **PRCLR** Pre-cooler
- **PRCS** Processor
- **PRECIP** Precipitation
- **PRED** Predicted
- **PREFLT** Preflight
- **PRELIM** Preliminary
- **PREP** Prepare
- **PRES** Present
- **PRESS** Pressure
- **PREV** Previous
- **PRF** Pulse Repetition Frequency
- **PRGM** Program
- **PRIM** Primary
- **PRL** Parallel
- **PROC** Procedure
- **PROF** Profile
- **PROG** Program
- **PROG** Progress
- **PROM** Programmable Read-Only Memory
- **PROT** Protection
- **PROV** Provisions
- **PROX** Proximity
- **PRR** Production Revision Record

- **PRR** Pulse Repetition Rate
- **PRSOV** Pressure Regulating Shutoff Valve
- **PRV** Pressure Reducing Valve
- **PRV** Pressure Relief Valve
- **P-S, PARA/SER** Parallel to Series
- **PS, P/S** Pitot/Static
- **PSA** Power Supply Assembly
- **PSCU** Programmable System Control Unit
- **PSE** Principal Structural Element
- **PSEU** Proximity Switch Electronic Unit
- **PSI** Pounds per Square Inch
- **PSIA** Pounds per Square Inch Absolute
- **PSID** Pounds per Square Inch Differential
- **PSIG** Pounds per Square Inch Gage
- **PSM** Programmable Switch Module
- **PSM** Power Supply Module (Monitor)
- **PSR** Primary Surveillance Radar
- **PSS** Passenger Service System
- **PSU** Passenger Service Unit
- **PSUD** Passenger Service Unit Decoder
- **PT** Total Pressure
- **PTC** Pack Temperature Controller
- **PTH** Path
- **PTO** Power Take-Off
- **PTR** Push To Reset
- **PTR** Printer
- **PTT** Press-To-Talk
- **PTT** Push-To-Talk
- **PTU** Power Transfer Unit
- **PVD** Para-visual Display
- **PVSCU** Programmable Video System Control Unit
- **PWR** Power
- **PWR SPLY** Power Supply
- **PWS** Predictive Wind-shear System
- **PYL** Pylon

Q

- **QAD** Quick Attach Detach
- **QAM** Quadrature Amplitude Modulation Unit
- **QAR** Quick Access Recorder
- **Q-CLB** Quiet Climb Thrust Mode
- **QDT** Quadrant
- **QEC** Quick Engine Change
- **QFE** Altimeter Setting To Show Altitude above Reference Airfield
- **QNH** Altimeter Setting To Show Altitude above Mean Sea Level
- **QTY** Quantity
- **QUAD** Quadrant

R

- **R** Right
- **R** Repeat
- **R** Range
- **R** Roll
- **RA** Radio Altimeter
- **RA** Radio Altitude
- **RA** Resolution Advisory
- **RA** Resolution Alert
- **RAAS** Runway Awareness and Advisory System
- **RABS** Reverse Actuated Bleed System
- **RAD** Radio
- **RAG** Repair Assessment Guideline
- **RAIM** Receiver Autonomous Integrity Monitor
- **RAM** Random Access Memory
- **RAM** Repair, Alteration and Modification
- **RAP** Repair Assessment Program
- **RAT** Ram Air Turbine
- **RAV** Ram Air Valve
- **RBL** Right Buttock Line
- **RC, R/C** Rate of Climb
- **RC, R/C** Rate of Change

- RCAS Roll Command Alerting System
- RCCB Remote Control Circuit Breaker
- R-CLB Reduced Thrust Climb
- RCDR Recorder
- RCL Recall
- RCM Ratio Change Module
- RCP Radio Communication (Control) Panel
- RCVR Receiver
- RCVR/XMTR Receiver Transmitter
- RDF Radio Direction Finder
- RDMI Radio Distance Magnetic Indicator
- RDNG Reading
- RDP Roller Drive Power
- RDR Radar
- RDR/XMTR Radar Transmitter
- RDS Radial Drive Shaft
- RDU Remote Display Unit
- RDY Ready
- REC Receive
- REC Recorder
- RECD Received
- RECIRC Recirculate, Recirculating
- RECP Receptacle
- RECT Rectifier
- RED Reduction
- REF Reference
- REFLD Reflected
- REFRIG Refrigeration
- REG Registry
- REG Regulate
- REL Relative
- REL Release
- REM Remove
- REP, REPEL Repellant
- REPR Reproducer
- REQ Requested
- REQD Required
- REST Restricted
- RET Retard
- RETR Retractable
- REU Remote Electronics Unit
- REV Reverse
- REV Reverser
- RF Radio Frequency
- RFLNG Refueling
- RGLTR Regulator
- RH Relative Humidity
- R/I Removal/Installation
- RIB Right Inboard
- RII Required Inspection Item
- RIP Repeat Item Program
- RIPS Recorder Independent Power Supply
- RJM Remote Jack Module
- RJU Remote Jack Unit
- RLS Remote Light Sensor
- RLY Relay
- RMCP Radio Management Control Panel
- RMI Radio Magnetic Indicator
- RMS Root Mean Square
- RMTE Remote
- RNP Required Navigation Performance
- RO, R/O Roll Out
- RO, R/O Repair Orders
- ROB Right Outboard
- ROM Read Only Memory
- ROT Rotation
- ROV Repair and Overhaul Vendors (Component Vendors)
- RP Radar Processor
- RPM Revolutions Per Minute
- RPS Revolutions Per Second
- RPTG Reporting
- RQMT Requirement
- RST Reset
- RSVD Reserved
- RSVR Reservoir
- RSVR Resolver
- RT Rate
- R/T Receiver/Transmitter
- R/T Reverse Thrust
- RTA Required Time of Arrival
- RTC Rudder Trim Control
- RTE Route
- RTL Ready to Load
- RTN Return

- R-TO Reduced Thrust Take Off
- RTO Refused (Rejected) Take-Off
- RTS Return To Service
- RTV Room Temperature Vulcanizing Rubber
- RUD Rudder
- RV Rated Voltage
- RV Relief Valve
- RVDT Rotary Variable Differential Transformer
- RVR Runway Visual Range
- RVSG Reversing
- RVSM Reduced Vertical Separation Minimums
- RVSR Reverser
- RVT Rotational Variable Transformer
- R/W Read/Write
- RW/ILS Runway ILS
- RWY Runway
- RX Receive

S

- S Second
- S South
- SAARU Standby Attitude/Air Data Reference Unit
- SAC Single Annular Combustor
- SAT Satellite
- SAT Static Air Temperature
- SATCOM Satellite Communication
- SB Service Bulletin
- S/B Speed Brake
- S/C Step Climb
- SC-A System Controller-Audio
- SCAV Scavenge
- SCF Standard Cubic Feet
- SCF System Card File
- SCHED Schedule
- SCM Spoiler Control Module
- SCR Silicon Controlled Rectifier,
- SCU Start Converter Unit
- SCU Seat Control Unit
- SCV Surge Control Valve
- SDI Source Destination Identifier
- SDR Service Difficulty Report
- SDU Satellite Data Unit
- SEB Seat Electronics Box
- SEB/ST Seat Electronics Box with Self Test
- SEC Second
- SEC Section
- SECT Section
- SEI Standby Engine Indicator (Instruments)
- SEL Select
- SEL Selector
- SELCAL Selective Calling
- SEL PNL Selector Panel
- SENS Sensor
- SENS Sensitivity
- SER Scheduled Engine Removal
- SERVO Servomechanism
- SEQ Sequence
- SEU Seat Electronics Unit
- SFC Specific Fuel Consumption
- SHP Shaft Horsepower
- SHT Sheet
- SID Standard Instrument Departure
- SIG Signal
- SIM Structural Inspection Manual
- SIS Service Interphone System
- SL Sync Lock
- SL Sea Level
- SL Service Letter
- SLCTD Selected
- SLCTR Selector
- SLS Side Lobe Suppression
- SLST Sea Level Static Thrust
- SM System Monitor
- SME Shipping Mechanical Equipment
- SMS Safety Management System
- SMYD Stall Management Yaw Damper
- SMYDC Stall Management Yaw Damper Computer
- SN Sign
- SN Serial Number

- **SNR** Signal to Noise Ratio
- **SNSR** Sensor
- **SO** Shop Order
- **S/O** Shut-Off
- **S/O** Standard Option
- **SOH** Start of Heading
- **SOL** Solenoid
- **SOLN** Solenoid
- **SOLV** Solenoid Valve
- **SOS** Save Our Spares
- **SOV** Shut-Off Valve
- **SPCU** Standby Power Control Unit
- **SP, SPD** Speed
- **SPD BK, SPD BRK, SPDBRK** Speed Brake
- **SP GR** Specific Gravity
- **SPI** Special Purpose Identification
- **SPKR** Speaker
- **SPL** Splice List
- **SPLR** Spoiler
- **SPLY** Supply
- **SPM** Surface Position Monitor
- **SPM** Standard Practice Manual
- **SPN** Spin
- **SPS** Standard Positioning Service
- **SPS** Single Point Sensor
- **SPSC** Single Point Sensor Card
- **SPU** Start Power Unit
- **SQ** Squelch
- **SQL** Squelch
- **SRA** Safety Risk Awareness
- **SRAM** Static Random Access Memory
- **SRCH** Search
- **SRM** Stabilizer Trim/Rudder Ratio Module
- **SRM** Structure Repair Manual
- **SS** Stick Shaker
- **SS** Single Shot
- **SSB** Single Side Band
- **SSB** Split System Breaker
- **SSEC** Static Source Error Correction
- **SSFDR** Solid State Flight Data Recorder
- **SSM** Sign Status Matrix
- **SSR** Secondary Surveillance Radar
- **SSSV** Solid State Stored Voice
- **STA** Station
- **STAB** Stabilizer
- **STAR** Standard Terminal Arrival Route
- **STBY** Standby
- **STC** Supplemental Type Certificate
- **STCM** Stabilizer Trim Control Module
- **STD** Standard
- **STE** Short Term Escalation
- **ST INV** Static Inverter
- **STRG** Steering
- **STRUCT** Structure
- **STS** Status
- **ST INV** Static Inverter
- **STWY** Stairway
- **SUBASSY** Subassembly
- **SUP** Support
- **SUP** Suspected Unapproved Part
- **SUP-NUM** Supernumerary
- **SUPPR** Suppression
- **SUPPR** Suppressor
- **SURF** Surface
- **SVCE, SERV** Service
- **SVO** Servo
- **SVU** Seat Video Unit
- **SW** Switch
- **SW** Software
- **SW** Special Work
- **SWA** Service Weight Adjustment
- **SWDL** Software Data Loader
- **SWL** Sidewall
- **SWS** Stall Warning System
- **SYM** Symbol
- **SYNC** Synchronize / Synchronization / Synchronizing / Synchronizer
- **SYN/DIG** Synchronous Digital
- **SYS** System

T

- **T** Time
- **T** Temperature
- **TACAN** Tactical Air Navigation
- **TACH** Tachometer

- **TAI** Thermal Anti-Ice
- **TAS** True Airspeed
- **TAT** Total Air Temperature
- **TBD** To Be Determined
- **TBV** Transient Bleed Valve
- **TBV** Turbine Bypass Valve
- **TC** Total Cycle
- **TC, TCCA** Transport Canada Civil Aviation
- **T/C** Top of Climb
- **T/C** Transit Check
- **TCA** Turbine Cooling Air
- **TCAS** Traffic Alert and Collision Avoidance System
- **TCC** Turbine Clearance Control
- **TCC** Turbine Case Cooling
- **TCF** Terrain Clearance Floor
- **TCU** Thermal Control Unit
- **TCV** Temperature Control Valve
- **TD** Time Delay
- **TD** Turbine Drive
- **T/D** Top of Descent
- **TE** Trailing Edge
- **TECH** Technical
- **TEL** Telephone
- **TEMP** Temperature
- **TERM** Terminal
- **TERM BLK** Terminal Block
- **TERMN** Termination
- **TFC** Total Flight Cycles
- **TFH** Total Flight Hours
- **TFR** Transfer
- **TGT** Turbine Gas Temperature
- **TGT** Target
- **TDL** Time Delay Logic
- **TDX** Torque Differential Transmitter
- **TDZE** Touchdown Zone Elevation
- **TE** Trailing Edge
- **TEMP** Temperature
- **TERR** Terrain
- **TFC** Traffic
- **TFC** Total Flight Cycle
- **TFR** Transfer
- **TGB** Transfer Gear Box
- **THDG** True Heading
- **THP** Thrust Horse Power
- **THR** Throttle
- **THR** Thrust
- **THR HLD** Throttle Hold
- **THR REF** Thrust Reference
- **THRM** Thermal
- **THROT** Throttle
- **THRSH** Threshold
- **THSHD** Threshold
- **TK** Track
- **T/L** Tilt
- **TL** Thrust
- **TLA** Thrust Lever Angle
- **TLR** Thrust Lever Angle Resolver
- **TLW** Time Limited Work
- **T/M** Torque Motor
- **TMA** Thrust Mode Annunciation
- **TMC** Thrust Management Computer
- **TMD** Thrust Mode Display (same as TMA)
- **TMS** Thrust Management System
- **T/O** Take-Off
- **TO** Turn-Off
- **TOD** Top of Descent
- **TO/GA** Takeoff/Go-Around
- **TPIS** Tire Pressure Indication System
- **TPMU** Tire Pressure Monitor Unit
- **TPR** Transponder
- **T/P Sensor** Temperature/Pressure Sensor
- **TQM** Total Quality Management
- **TR** Transformer Rectifier
- **TR** Torque Receiver
- **TR** Temporary Revision
- **T/R** Thrust Reverser
- **TRA** Thrust Resolver Angle
- **TRANS** Transition
- **TRANSF** Transformer
- **TRAX** Name of Company Computer System
- **TRC** Thermatic Rotor Control
- **TRF** Turbine Rear Frame
- **TRK** Track
- **TRU** Transformer Rectifier Unit
- **TRU** True
- **TS** Terminal Strip

- **TS** Temperature Switch
- **TSFC** Thrust Specific Fuel Consumption
- **TSLSV** Time Since Last Shop Visit
- **TSN** Time Since New
- **TSO** Time Since Overhaul
- **TSPR** Time Since Performance Restoration
- **TST** Test
- **TSTR** Transistor
- **TT** Total Time
- **TTG** Time to Go
- **TURB** Turbine
- **TURB** Turbulence
- **TURBO GEN** Turbine Generator
- **TVC** Turbine Vane Cooling
- **TVE** Total Vertical Error
- **TX** Transmit
- **TX** Torque Transmitter
- **TXPDR** Transponder
- **TYP** Typical
- **TYP** Type

U

- **U** Micro
- **UBR** Utility Bus Relay
- **UC** Upper Center
- **UCM** Un-commanded Motion
- **U/D** Upper Deck
- **UER** Unscheduled Engine Removal
- **UHF** Ultra High Frequency
- **ULB** Underwater Locator Beacon
- **ULD** Underwater Locating Device
- **UPR** Upper
- **USB** Upper Side Band
- **USB** Universal Serial Bus
- **USEC** Microseconds
- **USG** United States Gallons
- **UTC** Universal Time Coordinated
- **UTIL** Utility
- **UV** Ultraviolet

V

- **VHF** Very High Frequency
- **VIB** Vibration
- **VID** Video
- **VID** Valid
- **VIGV** Variable Inlet Guide Vane
- **VIS** Visual
- **VIU** Video Interface Unit
- **VLV** Valve
- **VMC** Visual Meteorological Conditions
- **Vmo** Velocity Maximum Operating
- **VNAV** Vertical Navigation
- **VO** Voice
- **VOL** Volume
- **VOR** VHF Omnidirectional Range, VHF Omni Range
- **VPLC** Valve Position Logic Card
- **VR** Video Reproducer
- **VR** Voice Recorder
- **VR** Voltage Regulator
- **VR** Rotation Speed
- **Vref** Reference Speed
- **Vrms** Voltage Root Mean Square
- **VRU** Video Reproducer Unit
- **V/S** Vertical Speed
- **VS** Video Switch
- **VSCU** Video System Control Unit
- **VSD** Vertical Situation Display
- **VSEB** Video Seat Electronics Box
- **VSF** Speed Floor Velocity
- **VSI** Vertical Speed Indicator
- **VSV** Variable Stator Vanes
- **VSWR** Voltage Standing Wave Ratio
- **VTK** Vertical Track Error
- **VTO** Volumetric Top-Off
- **VTR** Video Tape Reproducer
- **VTY** Vanity

W

- **W** West
- **W** Watt
- **W/A** Wrap Around

- WAF	Wafer
- WAI	Wing Anti-Ice
- WARN	Warning
- WBA	Wire Bundle Assembly
- WBC	Weigh and Balance Computer
- WBL	Wing Buttock Line
- WCP	Wing Chord Plane
- WDM	Wiring Diagram Manual
- WEA	Weather
- WEM	Warning Electronics Module
- WEU	Warning Electronic Unit
- WF	Weight of Fuel
- WG	Wing
- WG	Wave Guide
- WGT	Weight
- WH	Watthour
- WHCU	Window Heat Control Unit
- WHL	Wheel
- WICS	Work Instruction Card System
- WIU	Wire Integration Unit
- WL	Water Line
- WO	Work Order
- W/O	Without
- WOF	Whichever Occurs First
- WOL	Whichever Occurs Last
- WPT	Waypoint
- WRL	Wing Reference Line
- WRNG	Warning
- WS	Wing Station
- WSHLD	Windshield
- WT	Weight
- WTAI	Wing thermal anti-ice
- WTG	Waiting
- WTR	Water
- WTRIS	Wheel-To-Rudder Interconnect System
- W/W	Wheel Well
- WX	Weather
- WXR	Weather Radar
- WX/TURB	Weather/Turbulence

X

- X-By	Between Dimensions
- X-CH	Cross Channel
- XCHAN	Cross Channel
- XCVR	Transceiver
- XDCR	Transducer
- XFD	Cross-feed
- X-Feed	Cross-feed
- XFER	Transfer
- XFMR	Transformer
- XFR	Transfer
- XLTR	Translator
- XMISSION	Transmission
- XMIT	Transmit
- XMSN	Transmission
- XMTG	Transmitting
- XMTR	Transmitter
- XPC	External Power Contractor
- XPNDR	Transponder
- XTK	Cross Track (Deviation)
- X-Sec	Cross Section
- XTAL	Crystal
- XTK	Cross Track
- XTR	Cross Track

Y

- Y	Year
- Y	Yaw
- YD	Yard
- Y/D	Yaw Damper
- YDM	Yaw Damper Module
- YDS	Yaw Damper System
- YR	Year

Z

- Z	Zone
- Z	Zulu (→Universal Coordinated Time)
- ZFW	Zero Fuel Weight
- ZMU	Zone Management Unit
- ZTC	Zone Temperature Controller

Reference

- ASD-STE100, Simplified Technical English, ASD Association of Europe.
- 항공정비사 표준교재 항공정비일반, 국토교통부.
- ICAO ANNEX 1 Personnel Licensing, ICAO.
- ICAO ANNEX 6 Operation of Aircraft, ICAO.
- ICAO ANNEX 8 Airworthiness of Aircraft, ICAO.
- FAA 14 CFR Part 1 Definitions and Abbreviations, FAA.
- EASA Part 66 Aircraft Maintenance License(AML), EASA.
- Commercial Aircraft Maintenance Manual and Task Card.
- ChatGPT-4o, OpenAI.